A GAME PLAN
FOR SYSTEMS DEVELOPMENT

YOURDON PRESS COMPUTING SERIES
Ed Yourdon, *Advisor*

BENTON AND WEEKES Program It Right: A Structured Method in BASIC
BLOCK The Politics of Projects
BODDIE Crunch Mode: Building Effective Systems on a Tight Schedule
BRILL Building Controls into Structured Systems
BRILL Techniques of EDP Project Management: A Book of Readings
CONSTANTINE AND YOURDON Structured Design: Fundamentals of a Discipline of Computer Program and Systems Design
DE MARCO Concise Notes on Software Engineering
DE MARCO Controlling Software Projects: Management, Measurement, and Estimates
DE MARCO Structured Analysis and System Specification
DICKINSON Developing Structured Systems: A Methodology Using Structured Techniques
FLAVIN Fundamental Concepts in Information Modeling
FRANTZEN AND MCEVOY A Game Plan for Systems Development: Strategy and Steps for Designing Your Own System
HANSEN Up and Running: A Case Study of Successful Systems Development
KELLER Expert Systems Technology: Development and Application
KELLER The Practice of Structured Analysis: Exploding Myths
KING Creating Effective Software: Computer Program Design Using the Jackson Method
KING Current Practices in Software Development: A Guide to Successful Systems
MACDONALD Intuition to Implementation: Communicating About Systems Toward a Language of Structure in Data Processing System Development
MC MENAMIN AND PALMER Essential Systems Analysis
ORR Structured Systems Development
PAGE-JONES Practical Guide to Structured Systems Design, 2/E
PETERS Software Design: Methods and Techniques
ROESKE The Data Factory: Data Center Operations and Systems Development
SEMPREVIO Teams in Information Systems Development
SHLAER AND MELLOR Object-Oriented Systems Analysis: Modeling the World in Data
THOMSETT People and Project Management
WARD Systems Development Without Pain: A User's Guide to Modeling Organizational Patterns
WARD AND MELLOR Structured Development for Real-Time Systems, Volumes I, II, and III
WEAVER Using the Structured Techniques: A Case Study
WEINBERG Structured Analysis
WELLS A Structured Approach to Building Programs: BASIC
WELLS A Structured Approach to Building Programs: COBOL
WELLS A Structured Approach to Building Programs: Pascal
YOURDON Classics in Software Engineering
YOURDON Design of On-Line Computer Systems
YOURDON, LISTER, GANE, and SARSON Learning to Program in Structured Cobol, Parts 1 and 2
YOURDON Managing Structured Techniques, 3/E
YOURDON Managing the System Life Cycle, 2/E
YOURDON Structured Walkthroughs, 2/E
YOURDON Techniques of Program Structure and Design
YOURDON Writing of the Revolution: Selected Readings on Software Engineering
ZAHN C Notes: A Guide to the C Programming

A GAME PLAN
FOR SYSTEMS DEVELOPMENT

STRATEGY AND STEPS
FOR DESIGNING YOUR OWN SYSTEM

TROND FRANTZEN

KEN McEVOY

YOURDON PRESS
A Prentice Hall Company
Englewood Cliffs, New Jersey 07632

LIBRARY OF CONGRESS
Library of Congress Cataloging-in-Publication Data

Frantzen, Trond
 A game plan for systems development : strategy and steps for
designing your own system / Trond Frantzen, Ken McEvoy.
 p. cm. — (Yourdon Press computing series)
 Bibliography: p.
 Includes index.
 ISBN 0-13-346156-4
 1. Information storage and retrieval systems—Business. 2. System
analysis. I. McEvoy, Ken, 1948- . II. Title. III. Series.
HF5548.2.F733 1988
658.4'038—dc19 87-32624
  CIP

Editorial/production supervision
 and interior design: Elaine Lynch
Cover design: George Cornell
Manufacturing buyer: Lorraine Fumoso

Interior line art: Laurel Kennedy

© 1988 by Prentice-Hall, Inc.
A Division of Simon & Schuster
Englewood Cliffs, New Jersey 07632

This book can be made available to businesses
and organizations at a special discount when
ordered in large quantities. For more information
contact:

>Prentice Hall
>Special Sales and Markets
>College Division
>Englewood Cliffs, New Jersey 07632

All rights reserved. No part of this book may be
reproduced, in any form or by any means,
without permission in writing from the publisher.

Printed in the United States of America

10 9 8 7 6 5 4 3 2 1

ISBN 0-13-346156-4

PRENTICE-HALL INTERNATIONAL (UK) LIMITED, *London*
PRENTICE-HALL OF AUSTRALIA PTY. LIMITED, *Sydney*
PRENTICE-HALL CANADA INC., *Toronto*
PRENTICE-HALL HISPANOAMERICANA, S.A., *Mexico*
PRENTICE-HALL OF INDIA PRIVATE LIMITED, *New Delhi*
PRENTICE-HALL OF JAPAN, INC., *Tokyo*
SIMON & SCHUSTER ASIA PTE. LTD., *Singapore*
EDITORA PRENTICE-HALL DO BRASIL, LTDA., *Rio de Janeiro*

To our families, friends, colleagues, mentors, students, and critics—especially those listed below—thank you. We learned much from you.

Ian Chapman	Ray ("*Captain Midnight*") Deis
Tom DeMarco	Edsger Dijkstra
Linda Frantzen	Kirk Hansen
Lloyd Hiscock	Michael Jackson
Joan Lee	Bob MacDonald
Judi McCallum	Steve McMenamin
Judy Morrison	Mike Silves
Josie So	Rommie Vanderboor
Jerry Weinberg	Ed Yourdon

Trond Frantzen and Ken McEvoy
Toronto, Canada

CONTENTS

FOREWORD xi

PREFACE xiii

ABOUT THE AUTHORS xv

ACKNOWLEDGMENTS xvii

1 SYSTEMS DEVELOPMENT MODELS 1

    1.1   "All Things Flow . . ." 2
    1.2   Patterns of Business Activity 4
    1.3   The Problem with Systems Development 7
    1.4   A Game Plan for Systems Development 10
    1.5   Summary 12

2 A PRACTICAL STRATEGY FOR SYSTEMS DEVELOPMENT 14

    2.1   Strategy and Tactics 14
    2.2   Why a SuperSet Approach? 15
    2.3   "Lets Get Logical, Logical!!" 17
    2.4   Clients, Users, the Development Team, and the Dictionary 20
    2.5   Things Going on in the Background 23

2.6　The Guiding Principle: Adapt, Don't Adopt!　25
2.7　Summary　26

3　THE ESSENTIAL SYSTEMS DEVELOPMENT MODEL　27

3.1　The SuperSet System Development Model　27
3.2　"Ban Extraneous Documentation!"　30
3.3　The Context　30
3.4　The SuperSet Model's Process Abstracts　34
　　3.4.1　Set Up the Preliminary Agreement, 34
　　3.4.2　Create Essential Business Model, 37
　　3.4.3　Model System Functions, 43
　　3.4.4　Design the System, 46
　　3.4.5　Build and Test the System, 48
　　3.4.6　Conversion, 50
　　3.4.7　Develop Training Materials, 52
　　3.4.8　Implement the System, 53
　　3.4.9　Maintain the System, 54
3.5　Reflections on the Overview Context　57

4　THE SUBSTANCE OF THE MODEL　59

4.1　Introduction　59
4.2　The Detailed Diagrams　60
　　4.2.1　Set Up Preliminary Agreement, 61
　　4.2.2　Create the Essential Business Model, 62
　　4.2.3　Model the System Functions, 112
　　4.2.4　Design the System, 137
　　4.2.5　Build the System, 147
　　4.2.6　The Conversion Process, 152
　　4.2.7　Develop Training Materials, 152
　　4.2.8　Implementation, 155
　　4.2.9　Maintain the System, 157

5　IMPLEMENTING A SUPERSET APPROACH　175

5.1　Introduction　175

Contents ix

    5.2   Train Users in the SuperSet Model Concepts   175
    5.3   How It Was Done at MaxiMoney, Inc.   176
    5.4   Management Commitment—How Much?   179
    5.5   Emphasize Analysis and Design   179
    5.6   Maintain the SuperSet Model   180
    5.7   Implementation Paybacks   181
    5.8   Prototyping and the SuperSet Model   181

6 EFFECTIVE SYSTEMS TRAINING   185

    6.1   Objectives of Training   185
    6.2   Core Training   187
    6.3   Training for Management   188
    6.4   Training for Users   188
    6.5   Timeliness of Training   188
    6.6   Questions of Standardization   189
    6.7   Core Formal Training   190
    6.8   Skills Inventories   190
    6.9   Learning Through Mentors   191

7 MANAGER'S NOTEBOOK   193

    7.1   Do More Effective Walkthroughs   193
    7.2   Use Teams   194
    7.3   Conduct Formal Postproject Reviews   195
    7.4   Create Effective Documentation   195
    7.5   Project Management and Control   198
    7.6   Producers and Directors   203
    7.7   Automated Tools for Systems Development   205

8 SUMMARY   207

9 THE DICTIONARY   211

| 10 | THE DIAGRAMS OF THE SUPERSET MODEL | 264 |

BIBLIOGRAPHY 268

INDEX 273

FOREWORD

I have been dazzled by the array of new productivity tools and systematic approaches making their way into the information systems arena. The speed of release of new products continues to increase, and the spectrum of products aimed at information systems departments continues to broaden. New and more powerful tools to improve and automate systems development are providing answers to systems development problems.

Answers are not enough; solutions are needed. Whereas an answer provides a response to a problem, a solution provides the correct response to a problem. Too often only an answer is provided for a systems development problem because the structure of the answer is coupled to the development methodology rather than the problem.

This book is a clear, uncluttered guide to arriving at solutions to business systems development problems. It doesn't attempt to offer a definitive solution, but, rather, acts as a guide to finding different solutions for different problems. By focusing on the approach, GamePlan provides a framework for development that has the flexibility to provide solutions for unique problems, not just one answer.

Gord Lalonde
Chairman
School of Computer Studies
Sheridan College of Applied Arts and Technology
Toronto, Canada

PREFACE

Since you received this book, you have probably already scanned some of the pages, and before getting into any serious reading you are wondering about its contents. We imagine that you have thought of several questions already. Let's try to anticipate some of them.

Is This Book for Me?

> Maybe. If you are involved in any aspect of the process of system development, from beleaguered user to frustrated programmer, this book should give you some insight into what's been going on and how you can change it. If you've never run into problems, maybe "Poppy Pig's Garden" would be more appropriate.

Is This Another Book on Methodologies?

> No. A methodology is a pigeon hole into which one project might fit. This book is about cabinetry.

Why Should I Be Interested in a Game Plan for Systems Development?

> With the growing complexity of modern systems, we need all the help we can get. No two projects are the same; a strategy that helps us deal with anything we might come up against is invaluable.

But, Isn't This Just More Theory?

Yes. Until it's applied. Then it works.

But I Don't Use the "Structured Stuff"!

Not everybody does. Though we believe that the structured techniques are powerful tools in the system development process, this book isn't about structured techniques. It's about building systems—knowing how to start and what to do next.

Aren't There a Lot of Other Books Just Like This One?

Not that we're aware of. There are books about powerful tools and techniques, such as the various aspects of analysis and design. There are even books on methodologies. What we're trying to do, though, is to answer the question, "Where do I begin, and what do I do?" not by providing checklists, or specific tools, but by focusing attention on the nature of the problem and the environment in which it must be solved.

There's an Awfully Big "Dictionary" at the Back. What's It All About?

This book is an essential model of the system development process. The dictionary at the back is our data dictionary, defining the terms and concepts that are used on the models of the process of development.

So, What Won't This Book Teach Me?

Magic and witchcraft. COBOL. Where to buy things cheap.

Well, What Will It Teach Me?

With luck, and some application, we hope it will show you how to think about the art of systems development and how to focus on the problem to be solved.

About the Authors

Trond Frantzen graduated with a Bachelor of Arts in Business Education and Data Processing in 1968, while **Ken McEvoy** graduated with a Bachelor of Arts in Mathematics and Music in 1970. Ken is also a Fellow of the Life Management Institute (FLMI), *Data Processing* and *Pension Planning*.

Trond and Ken started their respective careers as programmers in the aerospace and financial services industries. Today they are principals in the consulting firm of Benetech Systems, Inc. of Toronto, Canada. They specialize in accelerated system development and advanced productivity techniques, with particular emphasis on rapid business system requirements definition, scope, feasibility, architecture, and user-sensitive analysis.

ACKNOWLEDGMENTS

We have been extraordinarily fortunate in our careers. While most system development professionals are battling just to get enough time and money built into the budget to accomplish the impossible, our clients and employers have consistently given us the freedom to grow by working with our ideas and concepts. Unlike the experience of many, our clients have agreed and allowed us to put our ideas in place, refine these new concepts, and develop workable, *commercially usable* systems.

They have understood and accepted that good, workable systems come from deeply involving the users and building the user's system, not constructing an unmaintainable but fascinating technological marvel. More importantly, our clients understood that the objective was never to build a system, but to solve a business problem. Accordingly, our commitment has been to solve those problems, not to impose an inappropriate technological rat's nest on their situation.

While many outstanding people have influenced our thinking, the real opportunities to apply our methods have been provided by our clients and employers. We thank them for giving us those opportunities. No one else could have provided us with the experience.

Finally, credit is due our colleagues and team members, who by working with us and using these methods to build real systems, have helped refine many of the details.

And while we were building those systems and writing this book, our families tolerated the funny hours we kept and the strange words we came to mumble in the middle of the night . . . ranting and raving about essential models and system blueprints. Thankfully, they put up with us, feigned understanding, and showed great love and patience while we forgot where we lived on occasion.

CHAPTER 1

SYSTEMS DEVELOPMENT MODELS

SUPERSET SYSTEM DEVELOPMENT MODEL

What Is It?

- A model of all the processes potentially involved in the development of a business system, whether manual, automated, or a combination thereof, from which a subset can be drawn to suitably describe any specific system's development process.
- A generic systems life cycle model showing the functions that will occur and the deliverables that must be produced in order to create the universal system.
- A strategy, or game plan, for systems development that suggests that each systems development process will be unique and that a unique model must be created for each system to be developed prior to its development.
- A book by Frantzen and McEvoy.

Who Might Need This?

- Users
- Project managers
- Data processing professionals
- Business systems analysts
- Those with questions to be answered
- Those with answers to be questioned

What's Inside?

- Data flow diagrams
- Process descriptions
- A dictionary defining potential deliverables
- Years of experience
- Wry humor
- Sage advice
- Heartwarming reassurance that it's possible to deliver systems that the user wants, on time, and at the right price

1.1 "ALL THINGS FLOW..."

We live in a world of change. The businesses we work for must constantly cope with changing customer demands, suppliers, material costs, competition, government regulations, and a host of other influences.

Similarly, the data processing environment is constantly developing, probably a lot faster than the businesses we're in. In the past this explosive change wasn't even an issue, since the DP environment was reasonably static. Systems hardware and software certainly changed, but the changes were minimal (one large mainframe replaces another; long live OS/360!). Tools and procedures were set in place for the management and implementation of data processing projects, and things progressed as well as we thought they could—even though the inflexibility of the tools often created more problems than they solved.

But now, with a trend toward ad hoc decentralization of data processing departments, the advent of powerful microcomputers, fourth and fifth generation languages, our comfortable, familiar world of systems development has been turned upside down—or perhaps returned to its rightful owners. Methods that appeared to work well for us in the past may not be a lot of help for building today's systems, or for the systems that need to be built tomorrow.

Specifically, many of us have used a *systems development methodology* or a *project life cycle* type of concept, commonly called SDLCs. Tools such as these eased the implementation of many DP projects for us in the past. These familiar tools, however, which describe the phases, processes, and deliverables of a DP project, may also begin to cause disasters.

It's possible that some data processing managers, encouraged by a past success with such a detailed methodology, may soon demand that every project done in their department proceed exactly the same way as the previous one, through the same step-by-step series of activities. This approach, while a valiant attempt at controlling the development process, will soon cause more problems and incur greater expense than having no project plan at all.

Robert Block identifies twelve categories into which he classifies the most common failures of projects. Not surprisingly, one of the twelve is "methodology failure":

> Methodology failures are failures to perform the activities needed to build the system: unnecessary activities may be performed, needed activities may be omitted, or activities may be performed incorrectly. Methodology failure may be due to the lack of a formal methodology as a guideline to the system builder, or to an overly rigid adherence to the adopted methodology.[1]

More recently, however, systems life cycle models have been created using the structured techniques. Some have been more strategic and general[2] than

---

[1] Robert Block, *The Politics of Projects* (New York: Yourdon Press, 1983), p. 4.

[2] Edward Yourdon, *Managing the System Life Cycle: A Software Development Overview* (New York: Yourdon Press, 1982).

other, more detailed versions.³ The introduction of these models (using the structured techniques) makes a lot of sense because a methodology is just a system used to develop some other kind of system. And since we now have a good set of tools, it seemed reasonable to use them to build our methodology.

So we did.

What this book suggests is that our new methodology, or "systems-building system," has to be created anew for each project. Just as no *one* computer system will meet the needs of all those who want computer systems, neither will *one* life cycle model satisfactorily yield a variety of different target systems for clients by rigidly following the same steps each time. In fact, the analogy between "packaged" SDLCs and "packaged" software is a pretty good one: in both cases the users tend to have only some of their needs met, and either frustration or a lot of modification has to follow.

1.2 PATTERNS OF BUSINESS ACTIVITY

Imagine a new employee arriving for work in the customer inquiry section of Amalgamated International Research, eager to get a foothold in the dynamic world of business after years of liberal arts.

Upon asking exactly how he is to deal with customer inquiries, his supervisor responds, "Oh, just any way you feel would be effective. Customer files are over there in those filing cabinets, if you think you need them. Just watch the others and maybe you can pick something up. You must have done a lot of work while you were in university; it's just like that, only more so."

Not only does the mind boggle that AIR could still be in business without any policies and procedures for dealing with customer inquiries, imagine how frustrating it would be for the new employee to try to function in such a nebulous environment. There could be no sense of satisfaction or accomplishment when there is no certainty as to what should be done, and no standards against which to measure progress.

Business activity—whether financial, retail, or manufacturing—consists of organized patterns of behavior.⁴ Raw materials arrive, are warehoused and entered in the Stock Inventory System. A mail order arrives and is sent to the Credit Verification Department. A customer calls to inquire about her account balance and the call is passed to, and handled by, the Customer Service Department.

³ Brian Dickinson, *Developing Structured Systems: A Methodology Using Structured Techniques* (New York: Yourdon Press, 1981).

⁴ Paul T. Ward, in *Systems Development Without Pain* (New York: Yourdon Press, 1984), discusses in detail how to model organizational patterns.

Sec. 1.2 Patterns of Business Activity

The important thing to notice is the presence of a pattern. Because there is a pattern, we know that every shipment of raw material, every mail order, and every client inquiry will be treated exactly the same way as the one before. Because the pattern is *predictive*, it can also be used to train staff, as well as being a guide to the staff to help them do their jobs, and a standard against which performance can be measured.

The collection of all patterns of activity within a company constitutes the company's policies and procedures (although not all of these procedures will be well documented). Without some organization to the activities at AIR, without some detectable, predictable pattern, staff morale and productivity and customer and shareholder interest will all rapidly dwindle to zero.

We've looked at a case where there were no evident patterns or procedures. Let's also consider the opposite extreme (it's only fair), in which a variety of tools and very specific procedures are provided. Imagine our new employee being told the following:

> All the employees of the Customer Service Department take elocution and speed-dialing classes. Moreover, at your desk you will have a copy of Watson's Business Thesaurus and a set of log and antilog tables. As for how to deal with any specific customer inquiry, we have the standard procedure document *"How to Use the New Tele-Phone as a Business Device."* We bought it from an independent business consultant some years ago. Follow it dogmatically; it hasn't failed us yet.

Our new employee isn't much better off, and it certainly isn't clear that the customers are going to get any better service either. If you've ever had

to deal with someone on the phone who is reading from a script, you'll know how alienating this can be for you as a customer, and how uninvolved and uncommitted the person reading the material can sound.

There appear to be plenty of tools available, though we're not sure exactly how useful they are. And the procedures document leaves a lot to be desired. It appears to have been written by someone who wasn't immediately familiar with the business objectives of Amalgamated International Research or the tools available to its employees. Also, it looks suspiciously like it has never been maintained or upgraded, although it's probably safe to assume that the scope and nature of the business have changed substantially over the years.

Even procedures developed in house, which may be very successful when first introduced, soon become stultifying and counterproductive when followed dogmatically and not altered as the business and its technology change.

This dogmatic, rigid procedural setting is no more satisfying to the employee or customer than the original chaotic one, and any business that's run on patterns this inflexible is bound to suffer at the hands of the competition. Since the real world—the one we live and work in—is very dynamic, then it stands to reason that business procedures and models must also be dynamic; a static procedure can only serve a stagnant business.

The ability to see patterns of business activity is the prime prerequisite for the successful analyst. The process of describing and documenting such patterns is analysis. But this book *isn't about analysis*, as important as that activity is to the development of systems. Many good people have written good works on how to do analysis. This book is about the *process* of developing business systems and a likely set of deliverables. Most solutions will include the automation of some subset of the system, though the approach we take will work equally well on a completely manual system.

All we need to agree on to start is that in any efficient and productive business, the activities that actually take place form patterns, and these patterns should be documented in some way so as to reflect the policies and procedures of the company.

1.3 THE PROBLEM WITH SYSTEMS DEVELOPMENT

The litany of problems surrounding systems development is one we are all too familiar with. Budget overruns, schedule slippages, project cancellations, user dissatisfaction—we've heard it all, over and over again.

In the past decade or so, the number and quality of analysis and design tools available to the systems builder has increased tremendously. Tactically, we're miles ahead. But strategically we haven't progressed very much at all.

The scenarios described for Amalgamated International Research seem pretty silly when pictured at the clerical business level, but their analogous situations are still far too common when dealing with systems development. What we mustn't forget is:

<u>The development of a business system is itself a business activity.</u>

A company that can't deliver its products won't survive. A company that can't develop business systems won't grow. Ask an executive what his company's objectives are, and the chances are that he will respond in terms of growth (i.e., increase 75 percent in market share in the Northeast and make an entry in the West Coast market). Delivery of products becomes a means to the end, which is growth. (Diversification shows how irrelevant any specific product can be.) Thus, for many companies, the development of business systems may be one of the most important activities in which they engage, more important, in fact, than the creation of any specific product.

So all those desirable features we seek in any normal business procedure we should also find in the procedure to develop systems.

In many data processing installations, employees follow a standard path of promotion and career development: programmer, analyst, project leader, manager, vice president of finance, and so on. At some point, an employee will find herself responsible for the creation of a new system. In more mature companies, it's usually a "rebuilding" affair. But often the only guidance offered is "You've worked on other systems here. You know what they look like. So get out there and build me one!" No pattern of activity is offered for this task, no procedure given.

With a bit of luck, however, and some money available in the budget, a variety of tools will be provided: courses in systems analysis and perhaps even structured design and programming. A library of all the latest books on hardware and software engineering is perhaps available. But the presence of tools does not give the answer to the question, "In this company, in this shop, what is the procedure for developing a system?"

Many of us have lived in, maintained, and even enhanced a variety of houses in our day; but given a limitless supply of lumber and tools, most of us still couldn't build one.

One pitfall associated with the lack of a clear set of procedures to be used in building systems is the tendency to focus on the target system as the *only* deliverable. It certainly is the key deliverable; it's what the users asked for, and if it doesn't appear, or doesn't work when it does appear, then the project has failed—quite miserably.

But there are other deliverables that are just as essential to system success. Trained users, quality users' manuals, systems documentation (which is certainly not entirely unimportant), testing plans, and operations manuals are among some of the other deliverables whose absence will eventually cause the failure of any new system. Without a specific systems development *model*, these other deliverables may get lost in the stampede toward the deadline. Any prepackaged methodology that doesn't lead to the production of the systems components and systems support documents that you need in your shop isn't worth the paper it's written on.

Sec. 1.3   The Problem with Systems Development

In fact, the more insidious problem is often this second scenario: the prepackaged methodology.

A variety of systems development methodologies are currently available. Some are in book format. Others require elaborate training sessions, acres of shelving for manuals, and vast infusions of cash. The problems are the same in either case: rigidity and irrelevance.

Let's, then, look at the characteristics of a successful business procedure. A good procedure should be: (PPF)

- *Predictive:* A procedure must explain what should be done in a given set of circumstances. It can thus be used both for training and as a guide for those doing the work.
- *Productive:* All activities of the procedure must contribute to the realization of the stated objective. Nonproductive (and certainly counterproductive) activities should not be included in a good procedure.
- *Flexible:* As the needs of the customers and the nature of the business change, the procedure must also adapt to the changing environment.

This list is by no means exhaustive, but already "packaged" methodologies run into problems with at least two of these requirements.

A universal systems development methodology, like the universal vegetable,[5] must be all the things to all people. It must contain in its procedures all the steps necessary for the creation of all kinds of systems. Even though certain steps could be made optional (e.g., "Do not produce screen design document 10B-X42.73 if there is no online component to the system"), there would be many mandatory deliverables in what is essentially a checklist-driven methodology. We have used such methodologies, and there are few experiences more frustrating than slaving for days or weeks over mammoth documents that "have to be produced," when you don't really know *why* they are being produced (except that they are on *The List*), and when you suspect that no one will read them.

One of the dangers of following any procedure mindlessly is the potential of coming to believe that filling out the forms and writing the required reports constitute doing the work.

Mindless adherence to a methodology creates the illusion that the development process is under control. It also transfers *responsibility* from people to paper. Sudden panic can set in among the project team as the deadline approaches, *even though the mandated deliverables are appearing on schedule*, because of a growing realization that something's missing, or

---

[5] "The Universal Vegetable, which will once and for all put an end to the ridiculous specialization of the turnip, the leek and the string bean, which will be at one and the same time bread, meat, wine and coffee, and yield with equal facility cotton, potassium, ivory and wool. The Universal Vegetable which Paracelsus could not, and Burbank dared not, imagine!" (*The Apollo of Bellac*, Jean Giraudoux)

that the team is "missing the boat." We call this the "[We] thought we knew, and we hadn't even shuffled the cards" syndrome.[6]

An "off-the-shelf" methodology may work wonderfully for one specific project, but it's like a snapshot of the pattern of activity that delivered that project. It's a static model of one specific development process. But no company can survive in a competitive, changing environment if it freezes its method for developing its products *or* for developing its business systems. What's really needed for any business procedure is a flexible, dynamic *model* of the systems building process that can alter and grow with changes in the tools and technologies used by the business.

1.4 A GAME PLAN FOR SYSTEMS DEVELOPMENT

So let's summarize. Businesses need procedures that reflect the patterns of activities that constitute their processing. These procedures must be, at least:

- Predictive
- Productive, and
- Flexible

The development of business systems is itself a business activity, and therefore we should have a procedure that reflects the activities necessary to create a business system. "Canned" methodologies are not suitable because they lack flexibility, and often contain activities that neither contribute to the business objectives nor resemble the standard administrative practices in your company.

So what's this book all about, if it's not another canned life cycle?

It's a *game plan* for systems development.

We must have procedures, yet no single procedure will do. The diversity of business systems that must be created is enormous: manual systems and automated; centralized and decentralized; batch, interactive, and real-time. Each kind of system needs a procedure we can follow to create it. But the nature of that procedure depends very much on the size, complexity, nature, and components of the system to be built *and on the tools used to create that system*. The major output of the systems development process is the new system. If we're not producing the same system over and over again, and if we have any faith at all in Louis Sullivan's dictum "form ever follows function,"[7] then we need a new procedure to develop each new system.

---

[6] From *Nightwood*, by Djuna Barnes (New York: New Directions Books, 1961).

[7] Louis Sullivan, "The Tall Office Building Artistically Considered," by Louis H. Sullivan, is an odd, beautifully written, and most interesting article that is surprisingly relevant

Sec. 1.4 A Game Plan for Systems Development

This book proposes a *game plan* approach to systems development that offers two major differences from traditional development life cycles:

- The *game plan* approach stresses the transient nature of any systems building model and the relationship between the model and the system it is being used to create. It encourages the creation of a new development model, or project life cycle, for each project, based on the system being developed and the environment in which the development is taking place.
- The use of a *"superset"* approach allows a model to be created that describes a broad range of processes and deliverables, from which a *working set* must be chosen for each project. The *SuperSet System Development Model* thus created will be different for each company, since it depends heavily on a company's environment and available tools. One of the first activities of any company adopting this approach would be to create its *own* SuperSet model. One of the first activities of any *development team* working on a project would be to select the *subset* of the model that would be a satisfactory procedure for the development of the system. Like any other system, the Superset model would have to be regularly maintained and enhanced.

The *game plan* is a strategy for developing procedures that can be used to develop systems. It is *not* a substitute for thinking or the development of good people. It is a system that can be used to develop systems that can be used to develop systems: a meta-meta system. (Like taking differential after differential of a quadratic equation, eventually we hope to be left with some constants.)

There. We've covered the difficult part. From here on in, it's all straightforward.

---

to modern systems engineers and architects. Sullivan discusses the "essence" of a problem, iteration, refinement, and how form (design or implementation) must, by natural law, follow functional definition (the essential business need) in architecture. He even discusses principles of normalization and canonical synthesis in the architectural world ("As I am here seeking not for an individual or special solution, but for a *true normal type*, the attention must be confined to those conditions that, in the main, are *constant* in all tall office buildings, and even mere incidental and accidental variation eliminated from the consideration, as *harmful to the clearness of the main inquiry.*"

The wonderful prose by Sullivan, and early foundation for what we call *event-based analysis* (which will be discussed in detail later in this book), deserves much more than passing mention. Visit your library and read the whole enjoyable article. It was first published in *Lippincott's*, pp. 403–409, March 1896; *Inland Architect & News Record*, pp. 32–34, May 1896; republished, with slight changes, in the "Architectural Discussion" department of *The Craftsman*, pp. 453–458, July 1905, under the title of "Form and Function Artistically Considered." In January 1922, *Western Architect*, pp. 3–11, republished the essay under the original title with the notation: "Mr. Sullivan himself states that he has nothing to add nor to subtract from his early statement."

1.5 SUMMARY

Businesses consist of organized patterns of activity. The documentation of these activities forms the procedures of the company. The process of systems development is a business activity, and procedures are needed to guide the process and those doing the work. Unlike customer account balance inquiries, no two systems being developed are the same, so a unique procedure is needed in each case. Canned methodologies will not suffice because they treat all systems the same and generally serve only to *"build stuff to hide behind."* Data processing personnel, especially those of the data-structured persuasion,[8] will realize that if the major output is different on each occasion, then the procedure for producing that output must also be different.

Project failure can often be seen to be primarily the result of a poor development methodology. One of the authors identified several criteria for a client in a 1983 study. Concerning methodologies, he reported:

> Successful development teams *adapted* methods, techniques and the life cycle methodology to their particular projects, rather than *adopting* the standards verbatim. In fact, the failure of one large project [over 350 person-months] can be attributed to high or dogmatic adherence to a particular structured meth-

---

[8] In very simple terms, "data-structured" designers suggest that the shape of a system will be isomorphic to the structure of the input (J.-D. Warnier), the output (Ken Orr), or the mapping of the input on the output (Michael Jackson). For a fairer and more complete explanation, see Kirk Hansen's *Data Structured Program Design* (Topeka, Kans.: Ken Orr and Associates, Inc., 1984) or any of the books by the above authors mentioned in the bibliography.

odology. They lost sight of their goals. . . . Adaptors, however, felt a sense of ownership and responsibility.[9]

The rest of this book is divided into nine additional major chapters.

Chapter 2, "A Practical Strategy for System Development," deals with the concept of a *game plan* and the *superset* approach as a strategy—why we should be doing it, who the players are, and basic guiding principles.

Chapter 3, "The Essential Systems Development Model," includes a review of a technologically neutral, *logical* high-level model for the development of systems, any system. It reviews the basic processes at a fairly high level of abstraction, and relates each process to other processes, or its neighbors. This section is a compact introduction to a practical development model.

Chapter 4, "The Substance of the Model," deals with one way in which this game plan for systems development can be incorporated into a working data processing environment, whether it be accountable to DP or directly to users. In this section a specific, detailed model is worked out, the *SuperSet System Development Model*.

Chapter 5, "Implementing a Superset Approach," discusses management's role in developing and implementing an internal SuperSet model. We also offer a number of suggestions on how to get it all working in your environment.

Chapter 6, "Effective Systems Training," presents a suggested training plan for systems developers, managers, users, and business analysts.

Chapter 7, "Manager's Notebook," includes additional discussions of project management, measurement and estimation, and suggestions of new ways to look at the development of systems.

Chapter 8, "Summary," does just that; and Chapter 9, "The Dictionary," follows with what is perhaps the book's most important section. The Dictionary offers a lightly written, conversational approach to definitions of project deliverables and terminology.

Finally, Chapter 10, "The Diagrams of the SuperSet System Development Model," provides a complete list of diagrams for easy reference.

The use of a superset approach to the creation of development models is, in our experience, the most cost-effective way of tailoring development models for individual projects. It incorporates a great deal of flexibility, without having to totally reinvent the wheel for each project.

The model we have built and present to you in this book is an abstract. Each reader and company *must* build their own, not use ours. The model you build should be built using the tools you use. We use Yourdon/DeMarco tools, so the model reflects that. Someone using straight Jackson could build a Superset using Jackson tools, but it would look and be different from the one we built.

[9] Trond Frantzen, "Productivity 83," MaxiMoney Inc., Toronto, Canada, March, 1983.

CHAPTER 2

A PRACTICAL STRATEGY FOR SYSTEMS DEVELOPMENT

2.1 STRATEGY AND TACTICS

Strategy refers to the approach used to realize a major objective—the over-all plan. In our situation, a procedure that describes and guides the development process is a strategic document.

Tactics refers to the individual elements of a plan—the specific activities that, one by one, move us closer to the realization of a major goal. The use of structured analysis, the size of the project team, and the use of particular hardware and software are all tactical decisions that, we sincerely hope, will move us most productively toward the delivery of the new system.

The choice of tools that we can use to create a model of the development process is a tactical decision. In fact, the use of the word *model* in the previous sentence is in itself a tactical suggestion, since we will use a modeling tool to help us build our development system.

Our task is primarily an analytical one. Just as an analyst creates models of systems, showing the processes that must take place to fulfill the *business* objectives for a specific system, we will build a model that will

Sec. 2.2 Why a Superset Approach?

show us what particular activities will allow *us* to fulfill *our* objective, the creation of a new system.

The analytical tool we will use is *structured analysis* as described by Tom DeMarco[1] and further refined by John Palmer and Steve McMenamin[2] and Paul Ward.[3] These authors and many others have stressed the advantage of using a graphic modeling tool rather than a narrative technique.

All the reasons these authors and others have given to justify the use of a modeling tool can also apply to the concept of modeling the development process. We need to know what we're going to do before we start to do it. We need to know how all the processes and tools fit together. Thus we need to create a model of the activities we will go through in building the new system. It's not adequate simply to begin by launching into a project and carrying on until we're in trouble; we must experiment with models of the development process that we will use. And through experimenting and adaptation we will finally arrive at a development process that best fits the particular requirements of our given project.

Even experienced analysts, designers, and programmers have a failure of confidence when first sitting down to a new project—that sinking feeling in the stomach that you don't know where to begin. It has sometimes been called the "blank page" syndrome.

The problem is that we don't acknowledge this crisis. We must recognize that the first task is *not* to jump into analysis *but to specify where to begin, and then what to do next, and then after that, and so on.*

In other words, we must have a map before we start out on a journey.

2.2 WHY A SUPERSET APPROACH?

Although we've been putting a lot of stress on the fluid nature of the business world we find ourselves in, many things will be constant from project to project within a given company or environment. These constant factors allow for the creation of a superset of activities that can be gathered together in a process model and used as a generic, or "first-cut," model of the development process. Such a superset would reflect the hardware, software, business terminology and activity, and any other characteristics of the group that created it. Some firms, for example, may use only one large mainframe computer. Administrative or audit procedures may be quite consistent from project to project. The nature of the corporate business activity will be much more similar between projects within a firm, as opposed to projects drawn

---

[1] Tom DeMarco, *Structured Analysis and System Specifications* (Englewood Cliffs, N.J.: Prentice-Hall, 1979).

[2] Stephen McMenamin and John Palmer, *Essential Systems Analysis* (New York: Yourdon Press, 1984).

[3] Paul T. Ward, *Systems Development Without Pain* (New York: Yourdon Press, 1984).

from different business sectors. Also, there will be many procedural similarities from company to company in the pulp and paper industry, but these procedures will be very different from the ones common to the insurance or banking sector.

In the later chapters of this book, a *SuperSet model* is illustrated that might be applicable to a general commercial environment. This is *not* a general superset (nor was it our intention that it should be); it is just a working example. Any book that tried to include a true and theoretically complete SuperSet of possible systems activities would be too heavy, and far too boring, to be of any real use. Moreover, it would be fixed in time and be out of date the moment it was published. Worst of all, it would rob the development team of the joy and responsibility of creating their own SuperSet and project development models. It's just not possible to find all the answers in a single book (not even this one).

We should also point out that we created this model using the Yourdon/DeMarco structured techniques because we have found them effective on other projects. Your *SuperSet System Development Model* should be built using the tools you will be using for analysis, and may as a result look quite different. Just remember that *your* SuperSet model must be built by the group using it, and that it must be constantly maintained and enhanced based on your experience in using it.

2.3 "LET'S GET LOGICAL, LOGICAL!!"[4]

DeMarco stresses the difference between the logical model of a system, which is implementation-*independent*, and the physical model, which is implementation-*dependent*.[5] McMenamin and Palmer introduce the concept of an "essential" model,[6] one that describes those activities that must take place for business to carry on. The *essential* model can be viewed as the solution model in a world in which perfect tools and technology are available. Of course, in our lifetime, we'll probably never have the opportunity to work in a world with perfect tools and technology, but it helps to think of it as such. It helps us to strip away the usual physical constraints and concentrate on the real essence of the problem. The solution we arrive at may be just a little bit better when we don't encumber the basics with all kinds of hardware, methods, and preferences that will date the solution and quickly become irrelevant as people and practices change.

In the case of systems development, the context diagram[7] of the essential model might well be as in Fig. 1.

The user (i.e., client) perceives a need for a problem. The problem won't go away, so she asks someone to provide a number of reasonable solutions for her. Systems developers confirm the substance of the user's business with the client and provide her with a number of alternative solutions. The client, after much gnawing, gnashing, and soul-searching, selects one of the alternatives and asks that someone build the appropriate system. The systems developers, using the company's data resource, eventually build a system that is delivered to the client. The client is happy.

Use of corporate resources (whether raw materials in inventory or supplies in the stock room) will be given to any area in a company that can demonstrate a need for such resources. This is true even in the case of resources with zero marginal cost (such as access to leased lines). We feel that the approach to a company's *data inventory* should be similar. In Fig. 1, the fact that "Company Data Inventory" and "The Client" are not directly connected (and appear at opposite ends of the diagram) does not mean they are unrelated—although it may appear that way in some companies. We do intend to imply, however, that the client does not have direct ownership of the data that is essential to the business; the data is owned by the company. Data is a very valuable resource, and should be held in inventory with as much control and understanding as a company would have of more conventional resources.

[4] Sir Isaac Newton-John, video and album, 1687.

[5] DeMarco, *Structured Analysis and System Specifications*.

[6] McMenamin and Palmer, *Essential Systems Analysis*.

[7] A context diagram is a single-bubble data flow diagram in which the one bubble represents the whole system. It is used primarily to illustrate the data flowing into and out from the system.

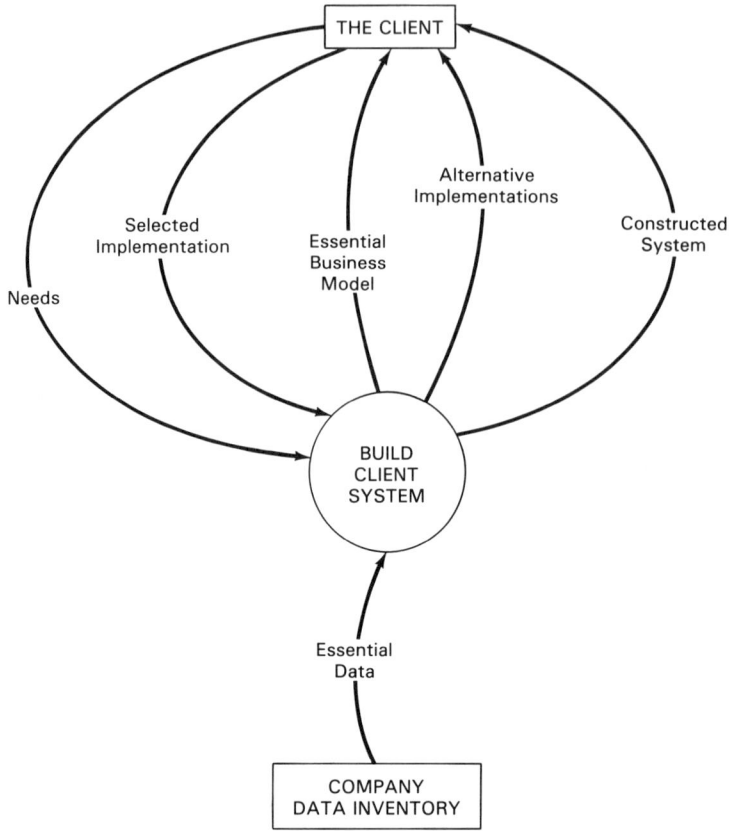

Figure 1: Context: Essence of Build Client System

When we look inside the context diagram in Fig. 2, we find the expanded processes that are essential to defining the client's new (or amended) system. (The idea of uncovering "essence" is to get to the innermost substance of the subject, without any encumbrances. Essential business models usually aren't very big.)

As we said earlier, the client realizes something needs fixing or enhancing. The nature of this pressing need doesn't particularly matter while we are addressing the fundamentals of what has to be done.

The client arranges for someone with suitable skills and credentials to take a look at her business; the systems developers gather some facts and define the essential business to uncover its innermost substance. This is sometimes called *function archeology*, since many layers of *civilization* can be uncovered before the real thing—essence, that is—may be found.

After confirming the fundamentals of the business and the problem with the client, they quickly offer her a number of choices. The client picks

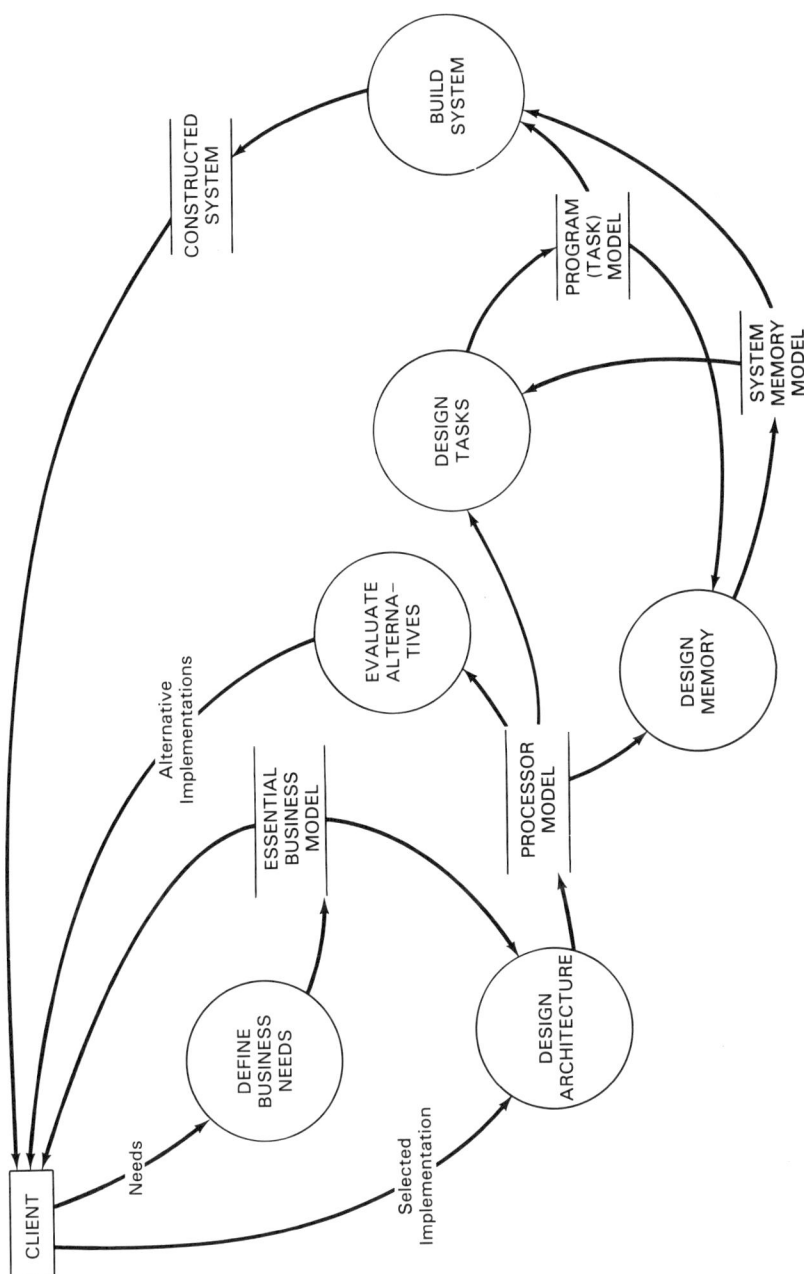

Figure 2: Essence of Build Client System—The Development Life Cycle

one of the alternatives (which may very well be to *not* solve the problem, in which case all action stops at this point), and authorizes a development team to implement a solution. Using a model of the essential system (in this case, business substance), the team builds a very specific, unambiguous model of all the things, or functions, that have to be included in the system the client wants.

Once all these functions have been specified (the *what*), the team proceeds to design the execution of those functions (*how* they will be carried out), and the memory the system will need (being in an imperfect world, we still need to record stuff somewhere). At this point the real system can be built. After picking up the right data from the company's data inventory, the developers can apply the data to the constructed system for delivery to the client.

In essence, this is all there is to solving any business systems problem and implementing the solution. It's also quite evident that this abstract model does not suggest any particular technology; memory, for example, could as easily be filing cabinets or index cards as it could be a sophisticated high-technology database. Nor does it get involved in any of the controls (such as project management deliverables) that are inevitably necessary in the real world of system building and problem solving.

Even at this level, though, some problems begin to develop. So we'll have to make some definitions and a few assumptions to make it easier to proceed.

2.4 CLIENTS, USERS, THE DEVELOPMENT TEAM, AND THE DICTIONARY

In Figs. 1 and 2, which define the substance of building a client's system, we include "The Client." If we're going to model a system that *develops systems*, then it seems it will have to deliver those systems as output to someone or something. A processor with no output seems somewhat sterile.

The client, then, is this external entity that triggers the development process and receives the solution system. In a more political sense, the client is the person or group with authority and money. It doesn't matter if you're in a traditional DP department writing software for a line area manager, working independently on a contract basis, or in the direct employ of the person who needs a system. The client is that person or group of people who commission and pay for the solution system. They are, however, usually in management positions.

Users, on the other hand, are those people who use the existing system and will use any new functionally equivalent system. Users exist at all levels of the corporate hierarchy.

"Clients," "users," and the "development team" are roles that must

Sec. 2.4 Clients, Users, the Development Team, and the Dictionary   21

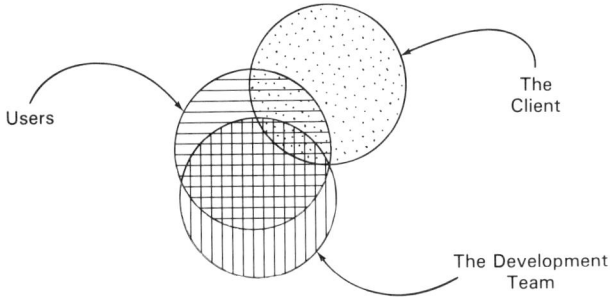

Figure 3: The Client Group, Users, and Development Team

be filled by members of the corporate population, which may include internal or external consultants.

See Fig. 3 for a diagram of the interaction among the client, the development team, and the users. In the process of creating models of systems, we talk about the *processes* taking place long before we consider the *processors* that will perform these functions. We may vaguely mention the word "processors" while making it clear to all our associates that we're not differentiating between manual and automated before the appropriate time. The *people* are the processors, though, in the process of systems development. In this book, the term "development team" is used to refer to all those people who perform any of the functions involved in the creation of a new system. It is irrelevant whether the development team consists of external consultants, procedure specialists, members of the data processing department, or members of the client's own staff. The latter is preferable from a practical point of view, because it is always the client who eventually pays the bill.

The members of the client group are usually a subset of the user community. The development team can be drawn from the user community as well (Fig. 3.1) or from external sources (Fig. 3.2), but it must include some representatives of the client and user areas. The arrangement shown in Fig. 3.3 is virtually impossible, and any project attempted this way is doomed to failure!

Systems developers must understand what they do, and why, so that quality systems can be delivered. Their basic business knowledge will be acquired when developing an **Essential Business Model**. But there is also a need to acquire an understanding of the detailed processes and handling of data that only the user can transfer. This intimate knowledge has to be reflected in the **Functional System Model**.[8] The only way to ensure the

---

[8] Occasionally you will encounter a word or a term set in boldface (when it is used for the first time.) This indicates that the term is explained further in the Dictionary (Chapter 9). As in any process model, the *data dictionary* is pretty important. It tells us what's what and avoids unintentional confusion. The same principle applies to this book and all the diagrams you will

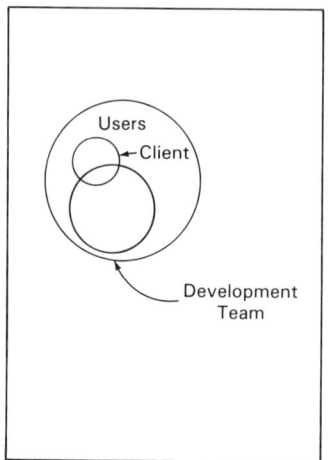
Figure 3.1

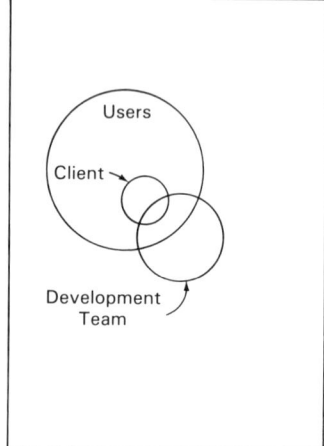
Figure 3.2

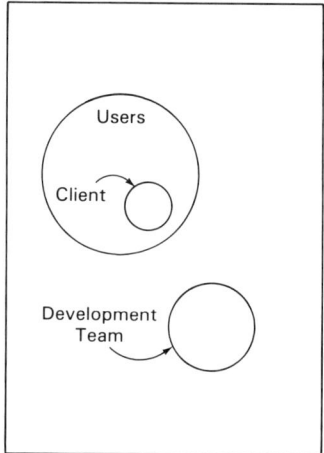
Figure 3.3

quality of these models is to get the user doing more of the business analysis and requirements work up front.

Users cannot be a group outside the project, but must be a committed, integral part of the **Development Team.**

In most companies the reality of the situation is that this isn't yet the case. So it is up to the data processing department to make the first move. In other words, put as much effort as possible into working more closely

find in it that are associated with our example SuperSet System Development Model. Because the Dictionary is so important to this book and its diagrams, use it all the time. Let it become your dearest friend. It will explain things, like data flows, or deliverables, or sources and destinations of deliverables, clearly and concisely. It's easy reading.

Sec. 2.5   Things Going On in the Background

with the users—so closely that they truly become "partners in development."

If DP remains isolated, user organizations will drift farther away from data processing groups than they already have. *Technology end runs* by the user community are already a reality in some organizations, and the price of technological error and omission is often steep. Isolationism has not worked well in politics, and has been hopelessly nonproductive for data processing groups. A participative development process, with the data processing department understanding that the client pays the bill, can lead to a partnership where quality and productivity are natural by-products.

Members of the Development Team are peers of the users and responsible to **The Client**, regardless of who their nominal managers are. As much as possible, we'll refrain from mentioning DP again. In fact, even in a very traditional DP development, certain activities (such as cost-benefit analyses) are much better carried out by the client's staff or users reporting to the client.

2.5 THINGS GOING ON IN THE BACKGROUND

There's another problem with explaining methodologies, including the one you're reading about now.

Imagine that you, as director of sales at Amalgamated International Research, decided that a series of full-page ads in *Scientific American* featuring a solution to Fermat's Last Theorem[9] would so impress the readers that AIR's market share would increase dramatically. So you hire a top-notch theoretical mathematician and assign her the task of coming up with a solution. Just to be safe, you also ask your best analyst to model the process of solving the problem: "It's just a one-person manual system, so put it on paper with bubbles and stuff!"

What would the context diagram of the essential process look like? Perhaps like Fig. 4.

So, what are the inputs? Not paper and pencil—that's too physical. Perhaps some reference texts? And how do we write the narrative that tells us what books must be read and when? Without further explanation, our diagram isn't exactly very meaningful.

The trouble here is that we can't model the *creative process* very well, if at all. In writing this book, we had the same problem when trying to show all the *stuff* that goes on in the background of the development process. Either we couldn't graphically express it with each set of diagrams, or the diagrams became so complex as to be totally incomprehensible. But there

---

[9] According to Fermat, there exist no positive integers x, y, z and n such that $x^n = y^n + z^n$. Unfortunately, he passed away before he was able to reveal his proof of this.

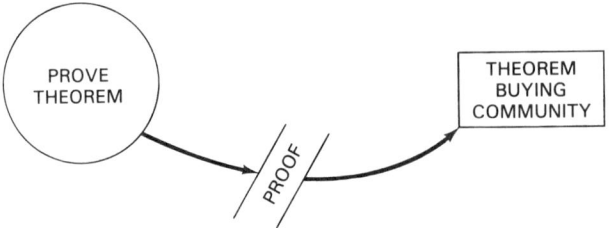

Figure 4: Context of the Process to Prove Fermat's Last Theorem

definitely is something going on behind every process, all the time. Graphically, it might look like Fig. 5.

Background research, thinking, and applied creativity are needed at all times to deliver the deliverables. Whether we need to build a business system or prove Fermat's theorem, the activity in the background is analogous. In terms of systems development, it's only because the process of building systems isn't entirely creative—because there are certain situa-

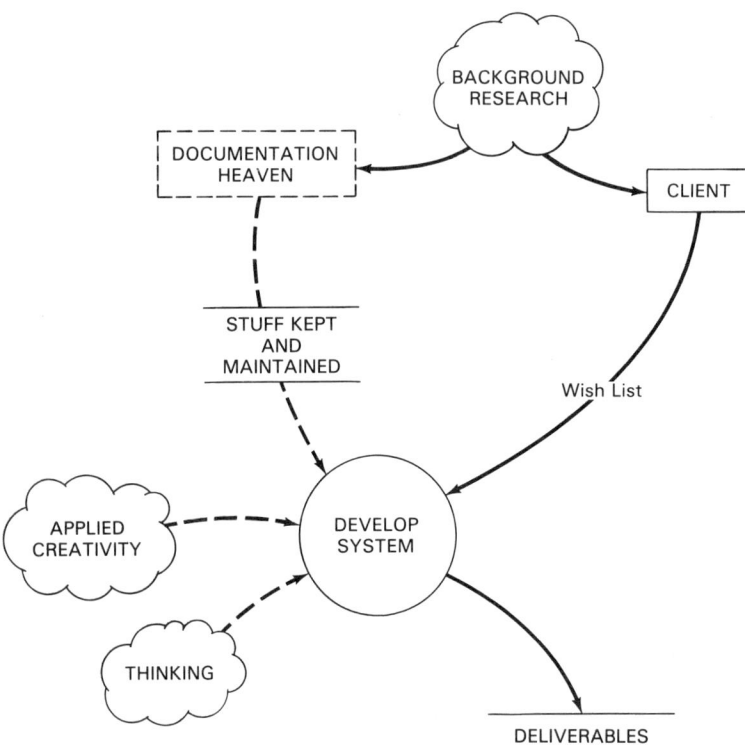

Figure 5: Stuff Going On in the Background

tions that recur from project to project, and similar inputs and outputs across development projects—that we can build any kind of model at all.

Also, if computers were not so absolutely noncreative (at least, so far), we wouldn't have to build any models at all. Not models of systems, nor systems-building models. But they are completely noncreative,[10] so, for the time being, we need systems models and models of the development process. And the easiest way to get a development model is to extract it from a "superset."

2.6 THE GUIDING PRINCIPLE: ADAPT, DON'T ADOPT!

It cannot be stressed too often that we must always be imaginative and flexible in our approach to problems. The world is always changing all around us. There are no high priests of systems development who have the right answer for all systems, for now, and for all time. In fact, not even for two of the above.

No life cycle approach should be *adopted*. A model of the development process *adapted* to your environment will work much more successfully. And even if you've had spectacular success with one project, don't try to use the same life cycle on another project: As project manager, you must readapt it before the project starts, based on the characteristics of the new system, its environment, and the changes that have taken place in your organization since you used it last.

The key deliverable is the implemented system. Not only must it be implemented, but it should be what the user wanted, on time, and within budget.

Always do the minimum work possible to reach these goals. Never follow an irrelevant standard. Don't permit yourself to spend any time on anything without fully understanding how the effort you're investing moves you closer to project completion.

Remember, productivity is maximized when only essential work is done.

---

[10] Well, they are getting more creative every day. Companies such as Syntelligence (based in Menlo Park, California) are developing and marketing expert systems that are powerful decision support systems which solve a variety of business problems through the application of new software technology. This technology enables computer systems to carry out judgmental analytical tasks that previously were the exclusive province of skilled professionals. Expert systems have been developed that exhibit reasoning capability at the level of the best human specialists in areas as complex as medical diagnosis, mineral exploration, chemical analysis, and computer configuration engineering. Most expert systems, however, currently are only capable of creating systems that are expert at identification and classification. The real breakthrough in artificial intelligence will come with new systems that are capable of performing tasks (i.e., doing things) rather than just glorified lookup procedures.

2.7 SUMMARY

The objective of a *game plan* approach to systems development using a *SuperSet model* is to produce systems faster, better, and cheaper. Members of the project team should always be identifying with the larger corporate objective of improved service, production, and profitability.

The use of a SuperSet Development Life Cycle allows the quick creation of a tailor-made process that will model the development of the new system. This project-specific process model will be the "road map" that shows how the various tools available can be integrated for the best benefit.

A company that wants to implement a superset approach, whether in a user or data processing department, will require, first and foremost, management commitment to the idea. Staff and management will need training in the techniques. So far, this is no different from acquiring *any* new systems development tool, such as a fourth generation language or Jackson design techniques. One of the most important steps in implementation, though, is the creation of a *SuperSet model* (or an adaptation of the one in this book) so that it reflects the business and administrative practices of the company, its existing and planned hardware and software, and the development tools currently available to it.

Again, *there is no single SuperSet model that will work for everyone.* We certainly don't pretend to be offering one. What we are offering is the concept that each company must create its own superset, using its own tools; thus each superset created this way will be unique.

CHAPTER 3

THE ESSENTIAL SYSTEMS DEVELOPMENT MODEL

3.1 THE SUPERSET SYSTEM DEVELOPMENT MODEL

A SuperSet System Development Model (see Fig. 6) is:

- a *superset* of the critical deliverables to be produced during the systems development process. Project teams are expected to select the appropriate deliverables for each unique project from this superset, and to modify and extend the resulting model based on specific project requirements in order to finally produce a "customized" life cycle model, unique to the project.
- a *network model*, or *map*, consisting of the paths (data flows) and processes that connect and create the superset deliverables.
- a *control mechanism*: the development model's "diagram 0" (system level) life cycle diagram provides the developer with a basic *game plan* for the development of a system. A developer is expected to decide which of the deliverables is appropriate for the project. As a management control check, appropriateness might be determined jointly by client management and the development team.
- a *documentation tool*: the deliverables produced through the development process (and none other!) constitute systems documentation.

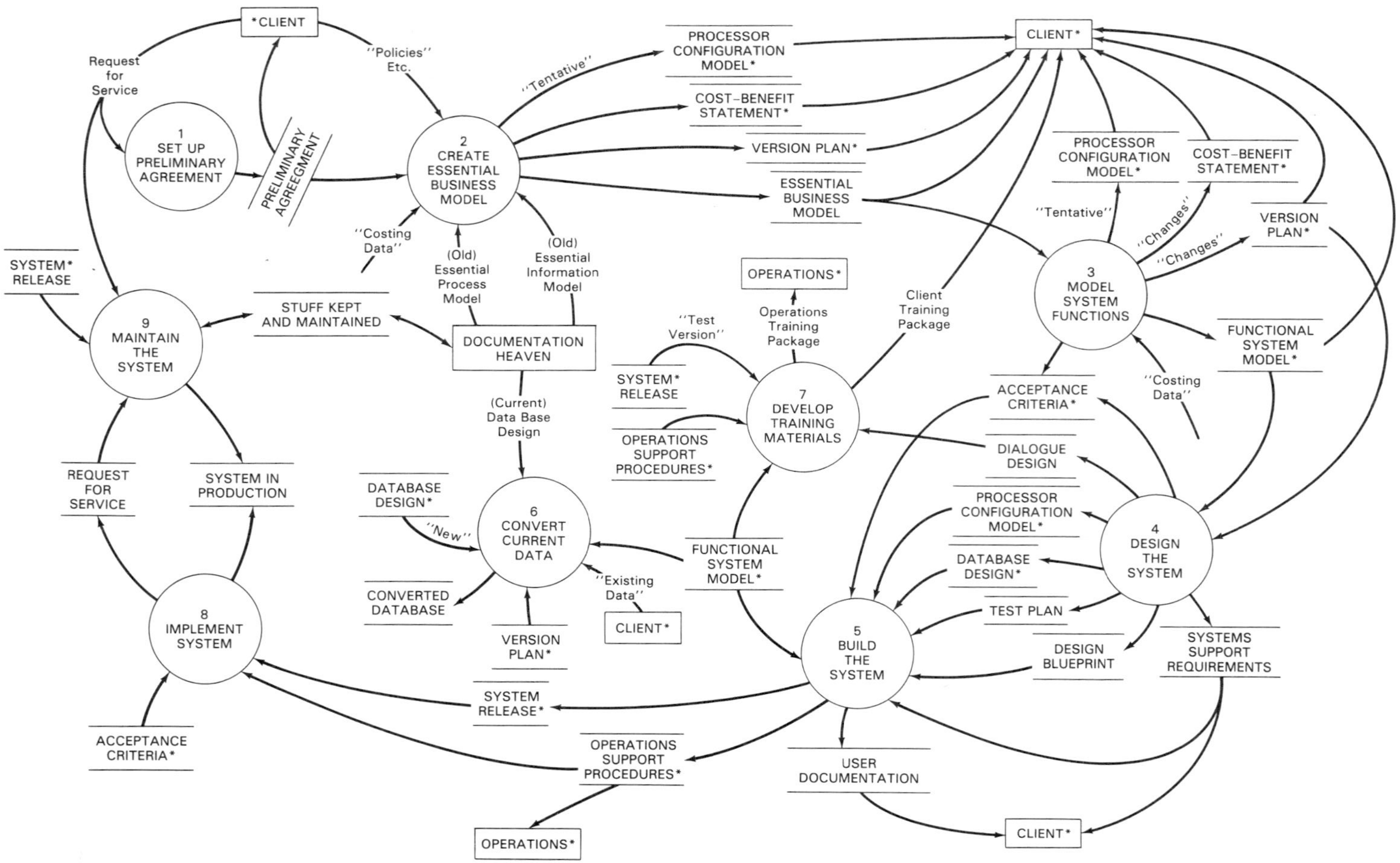

Figure 6: Diagram 0 of the SuperSet System Development Model

Sec. 3.1 The SuperSet System Development Model

- presented using *structured analysis techniques*: data flow diagrams, process specifications, and a data dictionary.
- defined, at the highest level, by *nine essential activities*: from project initiation, through two stages of analysis, then design, construction (and testing), conversion, training, implementation, and finally a somewhat different approach to the infamous maintenance ordeal.

Planning a major car trip without consulting a map is liable to lead to a lot of lost time, dead ends, frustrations, and constantly having to ask for directions. The use of a map to plan the trip will lead to much more efficient use of resources and will certainly be more enjoyable. But the map has to be up to date, with the latest highways and road closings. Using an out-of-date map is usually counterproductive.

The *SuperSet System Development Model* is a map for planning project development. It is not a detailed methodology. The finer points of how to do business systems analysis or systems design have already been explored far better than we can by a host of others referred to throughout this book. Rather, this is a *model* that reflects the experience of its creators. It's an example that reflects the experience of those who have done these things before, and should help you develop your own model. It makes alternatives available to development teams such that the leanest, most efficient project life cycle can be planned. Because the superset approach draws so heavily on what has gone before, the use of postmortems (a rather unfortunate term) or postproject reviews (which we will discuss later) becomes an essential part of the development process.

The SuperSet level 0 diagram includes nine activities, but these shouldn't be confused with "project phases." Too often, the concept of *phases* implies a serial development process. ("Well, phase 4 is finished! Let's have lunch, and then we can start on phase 5!") Each process in the level 0 diagram represents an activity that transforms input into output, and moves us closer to project completion. But like any data flow diagram, it's an asynchronous network model. Many of the activities may be going on simultaneously. In fact, on a project of any size, there will be very few times when all the efforts of the project team will be focused on one process. It happens, but it's pretty rare. Parts of the development model will be very much iterative, with activities being repeated over and over again, until we've refined it to the point of being satisfied with the result.

One thing you do not see on the system level (level 0) diagram is a process along the lines of "Create Systems Documentation." In fact, we suggest you shouldn't find this activity anywhere in the development model—not even at lower levels. There's a reason for this.

3.2 "BAN EXTRANEOUS DOCUMENTATION!"

After-the-fact documentation is seldom done, and less often done well. It's a loathsome task, often assigned to rookies or clerical level staff who haven't been involved with the system's development. It's often out of date before it's finished, as change requests for the system usually start to arrive long before the final documentation is tackled. And then, when the maintenance has to start, the fun begins.

Fortunately, the use of the SuperSet System Development Model gives us a neat out. By definition, there can be no post-development documentation, since the creation of all necessary documentation is part of the development process, regardless of when that documentation actually happens!

Moreover, it has been our experience that the documents produced through the use of current systems development tools (such as data flow diagrams, structure charts, entity relationship diagrams, data dictionaries, etc.), plus any user documentation needed, *together* constitute all the systems documentation needed. The Dictionary entry for Stuff Kept and Maintained shows that all the documentation produced by the systems developers is produced as part of the development cycle. The only exception to this rule of thumb is systems history, such as a change log.

If the users don't need it, if you didn't need it to build the system—for heaven's sake, don't write it!

If you are a project manager, or an analyst, or a designer, no doubt this philosophy will be near and dear to your heart. But it may not be entirely realistic in all environments. Some systems developers are contractually obligated to provide certain kinds of documentation. Government contracts are the most obvious example, and the Mil Spec documentation required by the U.S. Defense Department is probably the most onerous example. What do you do in this case? Well, you do the required extraneous documentation! Remember the philosophy: *if the users don't need it, if you didn't need it to build the system, don't write it!* In certain cases the users do need to meet a specified standard of documentation, or they can't obtain approval to continue with the building of the system, or it won't be accepted when implemented. Your only option in cases such as these is just do your best (life is hard), meet your obligations to the organization, and slowly try to bring about change. But if there is a choice, be the catalyst for change; *and if it won't move the project ahead, don't do it!*

3.3 THE CONTEXT

As illustrated in Fig. 7, *Development Life Cycle Context Diagram*, the input data flows are:

Sec. 3.3 The Context

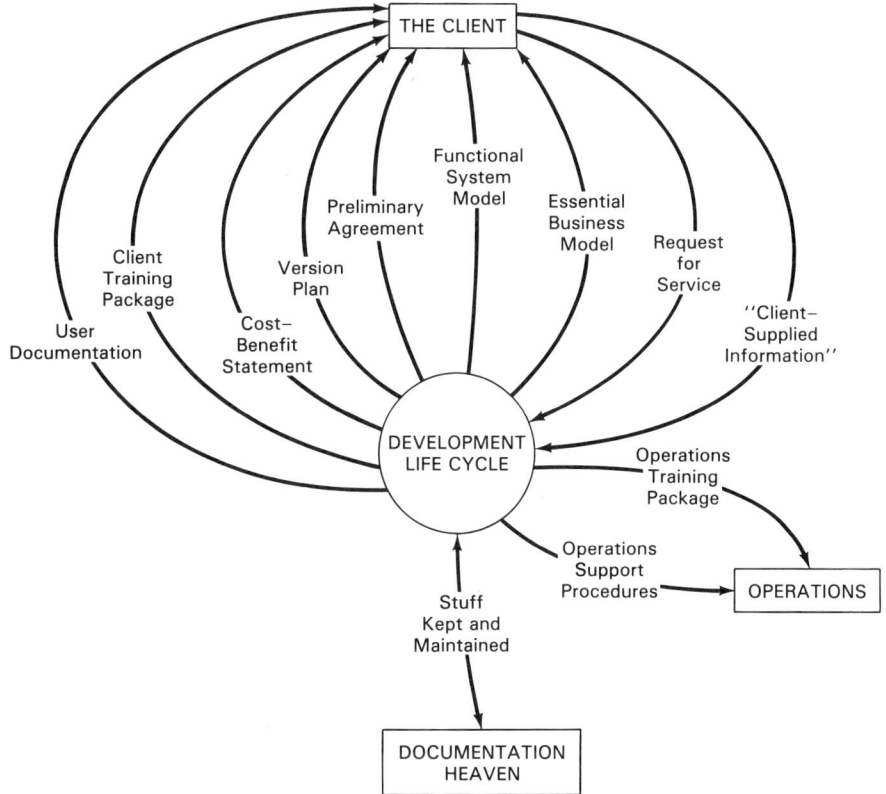

Figure 7: Development Life Cycle Context Diagram

- **Request for Service**
- "Client-Supplied Information"
- "Existing Documentation"

The output data flows are:

- **Preliminary Agreement**
- Essential Business Model
- **Version Plan**
- **Cost-Benefit Statement**
- Functional System Model
- User Documentation
- **Client Training Package**
- **Operations Training Package**

- **Operations Support Procedures**
- "Approved System Release"

The context diagram for the **SuperSet System Development Model** is very similar to the context for essence of build client system (Fig. 1). Some of the data flows have been given more detailed names than in Fig. 1. The content of these data flows are examined in detail as we work through the level 0 diagram in the next section.

One new terminator has been added: **Operations**. If the business systems we build will be at least partially automated, there will usually be an area of the company external to the client's area that will assume responsibility for providing the physical environment in which the system will run. It's also possible that there may not be an Operations Group in some companies. This could be because the group responsible for data processing is simply too small (the chief cook and bottlewasher idea), or because there are no mainframes, just a network of micro-computers. In fact, there may be no computers or automated portions to a company's systems at all. In this case, there will be no external support operation for computers. In any case, there is an analogy to groups such as plant or building services. This is the area of the company, external to the business systems, that controls the physical environment in which the systems get implemented, that is, the office. In some systems, the cooperation of such areas will be essential, and your working development model can be enhanced to include those extra activities. But in either case, there is the logical responsibility of Operations getting the work done.

The information gathered from the company's data inventory (in the Essential Model) will, in practice, either be interpreted for the Development Team by the client or will be found in real-life physical databases. The identification and gathering of essential data are therefore included inside the process of the development life cycle, that is, as part of *things going on in the background*. If your company has a separate group to do all data modeling and data administration, and the group is never part of the Development Team, then company data inventory would be a terminator in this context diagram. It isn't, however, because such a scenario is highly unlikely.

Since documentation is also important to the Development Team, when researching the existing system or when enhancements to a system are needed, there has to be a logical place to keep that documentation. Documentation is kept for one purpose only: knowledge. In real-life practice, the required knowledge is either interpreted for the Development Team by the users, or it will be found in **Documentation Heaven**, an external source of information needed by any project. Documentation Heaven is the aggregate of all information that should be recorded—corporate data policy, corporate business plans, models of existing systems, data dictionaries—all that neat stuff that we intend to leave behind when we finish a project. We've simply grouped this "documentation" under the name **Stuff Kept and Maintained**. This is all the stuff (not including miscellaneous deliverables needed by project management) that is normally delivered by the Development Team.

There are, however, two serious problems with Documentation Heaven:

1. It may not exist, in which case a lot of effort and expensive, valuable resources will have to be spent trying to gather the missing information.
2. The information present may be incorrect or out of date. If this is not recognized, the Development Team can go off on some very expensive side trips.

It's important to note that there is only one external event that triggers the process to start a project,[1] something similar to a "client wants system solution." The **Request for Service** is the stimulus to the systems development life cycle to start the process of building a system. Other stimuli (i.e., data flows) into the development process just represent the assembling of resources necessary to deliver the goods. One such data flow is "client-supplied information," which is a grouping for data relating to long-range plans, industry direction, legislation, and policies—all found at a lower level diagram. All of these data flows, coming from or going to terminators external to the systems development life cycle, are of course responding to additional "events." But these are all subsequent to the primary external event causing the project to start. These events only occur if and when a project to actually build something is authorized by the client. These would be events such as "operations group needs procedures manuals" or "management wants progress reports," most of which are driven by the management process. There is no doubt that there are several internal events, mostly temporal, but there is only one external event that counts and kicks off the project: the client wants something.

The rest of this section is structured very much like a walk-through. We'll quickly look at the level 0 diagram, discuss the processes and major data flows illustrated in that diagram, and give an overview of the life cycle.

Rather than trying to discuss the components of the diagram all at once, we felt it would be easier to stroll through each of the processes at the system level in turn, examining each in detail from two different perspectives. The diagram for each process is included as it is discussed. The

---

[1] Paul Ward discusses event modeling in *Systems Development Without Pain* (New York: Yourdon Press, 1984), pp. 149–158. Event modeling is perhaps one of the most powerful tools to be added to the analysis toolkit to date. We'll take a look at an example a little farther on, but a copy of Ward's book is an even better idea.

complete list of diagrams can be found in Chapter 10 and the Dictionary in Chapter 9.

3.4 THE SUPERSET MODEL'S PROCESS ABSTRACTS

In this section we will examine each of the development activities more closely.

Figure 8 presents diagram 0 of the SuperSet System Development Model again for easy reference. In order to look at all of these processes more closely, we create a context level diagram, or *neighborhood diagram*[2] for each activity, as if it were a complete system in its own right. The neighborhood diagram will show the process being discussed, the data flows entering and leaving the process, and the terminators of those data flows, though they will usually be other processes rather than true terminators.

This exercise—that is, looking at each individual process, its data flows, and its neighbors—will give us a better understanding of the data used in the process and *why* a process has certain neighbors. We will concentrate on the output side of each process.

3.4.1 Set Up the Preliminary Agreement (Fig. 9)

Purpose: To define the initial scope of the project; to identify who will pay for, approve, and accept the resulting system. In general, to "get the ball rolling."

                    Input: * Request for Service
                    Output: * Preliminary Agreement

The Request for Service is the only external data flow that triggers initial activity in the development system.

The Preliminary Agreement is a bit of a formality, but it does give some structure to what is usually a "vague mess." This document is essentially needed to clarify:

- who the client and users are
- who the approval body is
- what the scope of the project is
- what the major objectives of the client might be
- what major constraints there may be on any possible solutions

---

[2] Almost everything we talk about now is a model of one kind or another. We wanted to stress the similarity between our neighborhood diagram and a context diagram, while at the same time not add to the general confusion of terms by calling it another kind of model.

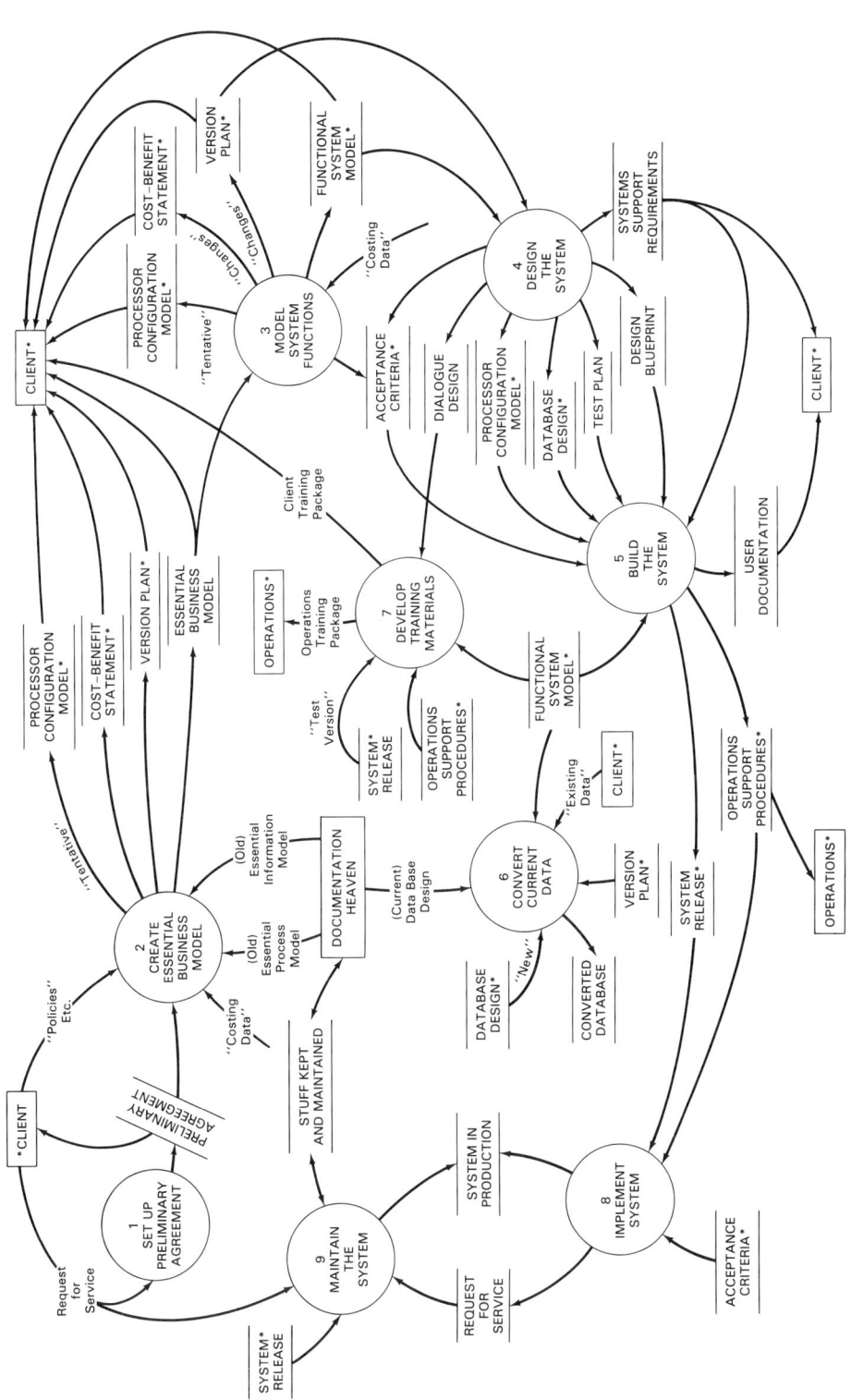

Figure 8: Diagram 0 of the SuperSet System Development Model

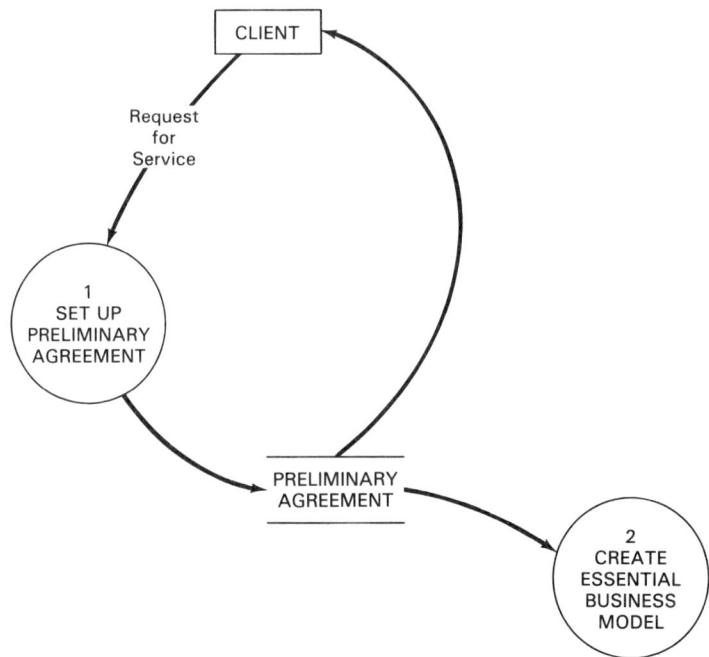

Figure 9: Neighborhood Diagram—Set Up Preliminary Agreement

The Preliminary Agreement shouldn't be labored over. Two or three days is usually sufficient, sometimes even less. Even the Preliminary Agreement for the largest, most complex projects shouldn't take more than a week. The Preliminary Agreement can be viewed as a contract between the client and the Development Team; in fact, the Preliminary Agreement should be much like the sort of contract you would draw up if you were engaging a consulting firm on a time and materials basis. It's just an agreement to start work, specify who does what, for whom, what the major deliverables are, what the expectations are, and who gets to make approvals. With this basic information as a guideline, the Development Team can start to develop the Essential Business Model.

This sounds really easy, and it is, but the project manager must do some advance work first! There are issues to be addressed and resolved that do not really get done when preparing a simple Preliminary Agreement, but must be agreed to by all participants in the development process—elements such as determining how the project will proceed, who is involved, understanding the skill level of available team members, assessing training needs, and so forth. All of these factors will affect results and must be written into the **Objectives & Constraints** section of the Preliminary Agreement to avoid unrealistic expectations.

3.4.2 Create Essential Business Model (Fig. 10)

Purpose: To determine if the client's problem can be solved; if solving the problem fits into long-range corporate plans; and to ascertain if a new business system can be built.[3] Given positive answers to these questions, the Development Team has the responsibility of delivering an Essential Business Model to the client. This is a basic model of the client's department, section, or business handling unit.

The Essential Business Model (EBM) consists of two major components: the **Essential Information Model** (EIM) and the **Essential Process Model** (EPM).

The Essential Information Model documents the data and the relation-

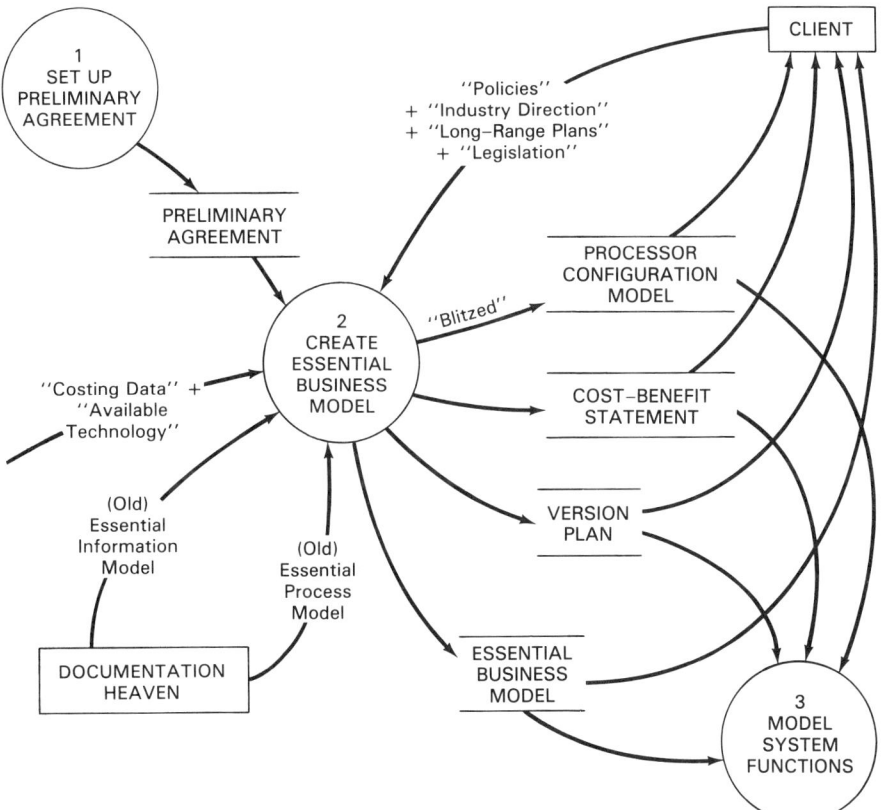

Figure 10: Neighborhood Diagram—Create Essential Business Model

---

[3] There's no suggestion at this point whether any part of the system should be automated or not. The Development Team probably started out with some ideas about what could be automated, but it shouldn't influence the Essential Business Model.

ships among the data (the business knowledge of the client), and also the policies behind the knowledge that guide the conduct of the business—what the client and users are allowed (or supposed) to do.

The Essential Process Model illustrates what is done with the data to achieve the desired objective—the "how" of the business.

Together as the Essential Business Model, the EIM and the EPM answer the questions:

> What are those things that we must do to continue in business? What are those activities that are essential in reaching our objectives, and what are the things we must remember in order to carry on with these activities?

The Essential Business Model has few, if any, physical constraints. Nor does it contain anything related to issues of control. It is the innermost substance of the client's business. It should be fairly simple when complete, and since it reflects the heart of the business, it shouldn't change very much over time. Any differences that do arise reflect a change in the nature of the business, and it is important that the client understand and agree that this change is an accurate reflection of what has actually happened.

Every project team must create (or revise) the Essential Business Model as a first step.

Input:
* Preliminary Agreement
* (Old) Essential Information Model
* (Old) Essential Process Model
* "Client-Supplied Information"
* "Costing Data"

Output:
* Essential Business Model

  > Essential Process Model
  > Essential Information Model

* Version Plan
* Cost-Benefit Statement
* "Blitzed" **Processor Configuration Model**

The Essential Business Model is a *logical* model: it must describe the data and processes necessary to do what the client wants, without any consideration of technology or the specific elements that will be used to implement the system. In reality, though, the client will always want to have an estimate of the duration and cost of the project under consideration, as well as a statement that can be used to compare alternative solutions. For this reason it is necessary, even at this early stage, to put very tentative automation and processor boundaries on the Essential Process Model and to give some consideration to possible implementations. When the Essential

Sec. 3.4 The SuperSet Model's Process Abstracts

Process Model has various automation and processor boundaries applied to it, including processes to be carried out by people, we simply call it the Processor Configuration Model. It's crucial, though, that both the client and the Development Team realize that this conjecturing is for cost-estimating purposes only and should not in any way be a constraint on the rest of the project.

On the other hand, there may be some very physical constraints that are completely outside the control of both the client and the Development Team. A typical example is the necessity of receiving or delivering data in a currently specified format, perhaps using a specific technology (usually old):

> Each box of parts from the Terre Haute factory comes with an envelope containing a punched card describing the contents and with some cost data. Not only do we need these data, but we also have to telex the boss in Pago Pago if there's been an increase in cost.

Such existing physical constraints, regrettable as they may be, can't (and shouldn't) be denied. If it is essential to the business to work in a certain physical way, then that should be reflected in the Essential Business Model. These physical constraints will normally be on the boundary of the system, though, and shouldn't be confused with the existing physical process. In other words, you don't need to anticipate using a DataSchwartz 2398 in the future simply because it's the only one you have now; all options are open to you when creating the Essential Business Model, and they should remain open as long as possible.

In the process of creating a cost estimate for the client, based on an Essential Business Model, it may be possible to describe a series of projects, or deliverables, that will gradually improve the procedures and solve the problem that brought the client to you in the first place. Such a delivery schedule, or Version Plan, is based on the tentative Processor Configuration Model (which reflects all the essential processes) and will require certain assumptions or decisions about the eventual implementation of the system. Depending on the system, versioning[4] such as this can also provide sufficient lead time to purchase computer or office equipment. Although this kind of versioning is but another example of physical constraints intruding too early, there is a payoff: it is planned, and the resulting implementation can be packaged into a series of projects, each of which is a manageable size. Moreover, the client will get some improvements more quickly, and the feedback from this will facilitate later versions of the system.

The Essential Business Model is needed by the Development Team in the next process (*Model System Functions*) so they can expand the models of the client's business into a complete specification. They will need the

---

[4] "Versioning" works especially well if "verbifying" is part of your corporate linguistic strategy.

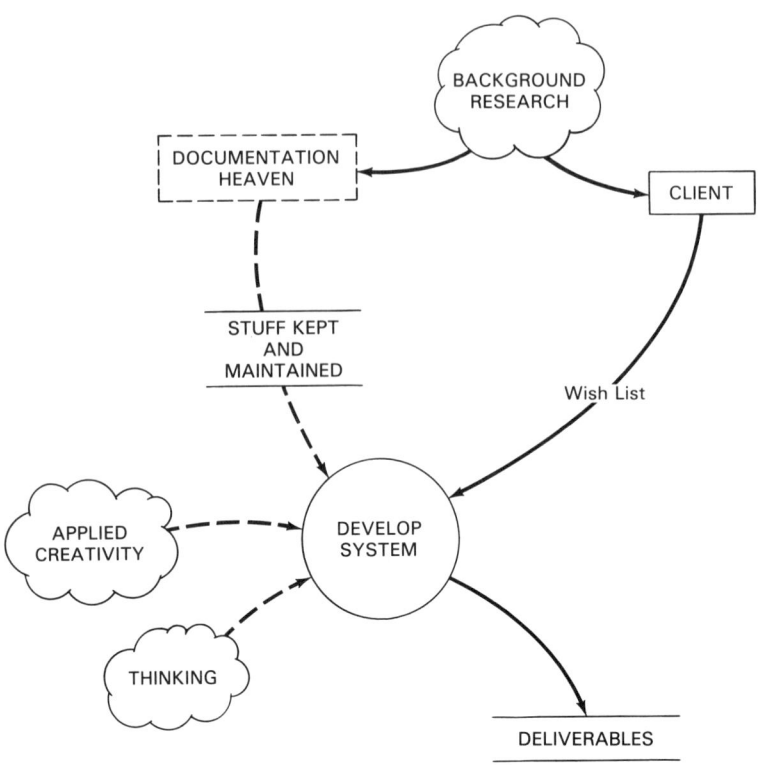

Figure 11: Stuff Going On in the Background

Version Plan so that they can concentrate only on the functions to be included in the current version and yet still be able to see the long-range plan for development. The Cost-Benefit Statement will be further refined as the project continues.

Before proceeding further, let's again briefly discuss the Development Team and what goes on in the background of the development process.

Because the Development Team consists of a variety of random, often eccentric, types, each working in his or her own unique way, it's not quite possible to create a completely rigorous and specific model of the development process. There's always stuff going on in the background—see Fig. 11.

Remember this picture? Well, throughout the development process, members of the Development Team will be doing research, talking to the client, the users, other employees, and so on. Much of this material will be gathered through informal interviews and meetings and not be put to paper in any specific form until it forms a cohesive unit. It will be collected in their minds, and perhaps in obscure personal notes. In addition to normal fact gathering, a whole lot of synthesis that is dependent on the previous experience of the members of the Development Team will go on to produce the Information and Process models that will define the client's system.

Sec. 3.4 The SuperSet Model's Process Abstracts   41

Although the definition of the various deliverables of the development process can be quite explicit, the creative process of producing the deliverables is still very much an art rather than a science. Almost anyone can be taught the basic rules of drawing, yet few people can become good artists. Good analysts often appear to have some mysterious ingredients that we can only refer to as the *right stuff*.

This basic activity—the research, the synthesis, the creativity—is going on in the background all the time. As you look at the process diagrams in this book, mentally superimpose the "background" picture on each of them. It may answer some of the questions you'll have and make it all seem less magical.

During the time the Development Team is modeling the substance of the system, however, they will constantly have in the back of their minds a lot of "physical" considerations, such as:

- "terminator" requirements
- data processing guidelines
- conversion considerations
- financial constraints
- hardware considerations
- existing commercial software
- physical plant or office
- scheduling and resources
- systems software constraints (the impact of operating systems, database management systems, etc.)

E. W. Dijkstra made this point well, as long ago as 1974 when discussing the same issue with Donald E. Knuth, as Knuth reported it:

> I feel somewhat guilty when I have suggested that . . . (you can) think about only one level at a time, ignoring completely the other levels. This is not true. You are trying to organize your thoughts; that is, you are seeking to arrange matters in such a way that you can concentrate on some portion, say with 90% of your conscious thinking, while the rest is temporarily moved away somewhat towards the background of your mind. But that is something quite different from "ignoring completely": . . . You remain alert for little red lamps that suddenly start flickering in the corners of your eye.[5]

Just as with Dijkstra talking about the impossibility of focusing on one "level of abstraction" only, the Development Team will always be concerned with these issues but few will get documented until they are finalized,

---

[5] E. W. Dijkstra, in Donald E. Knuth, "Structured Programming with Go to Statements," *ACM Computing Surveys*, December 1974, p. 308 in *Classics in Software Engineering*, Edward Yourdon, ed. (New York: Yourdon Press, 1979).

much later. All of this is stuff going on in the background and defies modeling.

Making sure that these things are adequately considered early enough in the development process, yet without building in unnecessary technological constraints, is very much the responsibility of the project manager.

In other words, the project manager must:

1. understand the deliverables of the development process;
2. fully comprehend the asynchronous, very iterative nature of the development process; and
3. choose the right people with the "right stuff" to do the development.

An inexperienced project manager, or one who does not fully understand the "deeply intertwingled" nature of developing systems may create many serious problems by premature consideration of those physical features that eventually have to be built into every system.

Traditional systems development methods are based on a very procedural and rigid approach. As a result, they focus on having clearly defined requirements and tasks very early in the process, minutely detailed plans, and rigid milestones. A group of people are then assigned to execute the individual tasks. "Individual tasks" usually means teams of one working in a corner—in short: a linear and rigid approach with a standard team.

On the other hand, the reality of modern systems development, as we understand and have experienced it, dictates a somewhat less structured approach. In the new systems development world—rather than the artificial world imposed by highly "structured" third generation techniques and project management methodologies—users have, at least initially, relatively undefined requirements in a changing business environment. They are also learning about the magnitude of their needs as those needs are specified. We therefore need a very structured and skilled team, with each player's role being clearly defined and each member of the team being an "expert" in his or her own right—in short: a structured team with an unstructured approach. An unstructured approach simply means working with the knowledge that we won't get it right the first time out. This is what structured analysis and design are all about: being experts at refining through iteration until we get it right. It's another form of prototyping, but quite explicitly takes into consideration the bigger picture rather than just a suite of programs.

Because of the highly iterative nature of the modern, fourth generation development process, changes in previous perceptions will cause more change again. This approach (which is reality, anyway) will usually make a shambles of a detailed task list that is produced too early. Traditional elaborate plans[6] in such an environment are not very practical because the

---

[6] "Appropriate schedules" come from Project Management Heaven.

tasks, and therefore the plans, are always changing. If progress, then, seems to be somewhat unpredictable ("unstructured") on a day-to-day basis, have another look at the team; it probably isn't organized, structured, or well enough skilled to do the specific job.

3.4.3 Model System Functions (Fig. 12)

Purpose: To expand the Essential Business Model into a complete, real-world specification statement. The primary deliverable, the Functional System Model, will illustrate and specify all the stimuli and data essential to the processes that deliver the planned responses of the client's system.

In this process the Development Team will build into the target system's model all the necessary control functions, such as edits, audits, and approvals. Processes that transport data (files, reports, or telecommunications links) will also be added, as will any processes necessitated by the known technology to be used.

Input: * Essential Business Model

          > Essential Process Model
          > Essential Information Model

      * "Blitzed" Processor Configuration Model
      * "Costing Data"

Output: * Functional System Model

          > **Process Model**
          > **Information Model**

      * "Tentative" Processor Configuration Model (revised)

      * **Acceptance Criteria**
      * (Revised) Cost-Benefit Statement
      * (Revised) Version Plan

The process of expanding the model of the system's functions at this point is very similar to the previous step of creating the Essential Business Model. The inputs and outputs of the process are also very similar.

There are two key differences.

First, the client has now made a strategic decision as to how a solution to the business problem can be realized. If the client chooses to do nothing (i.e., "maintain the status quo," one of the alternatives available), then the project is concluded. If any other choice is made, however, the project will carry on. If the chosen path is to develop a new system, the processes of developing the new system can be fairly easily defined using the superset

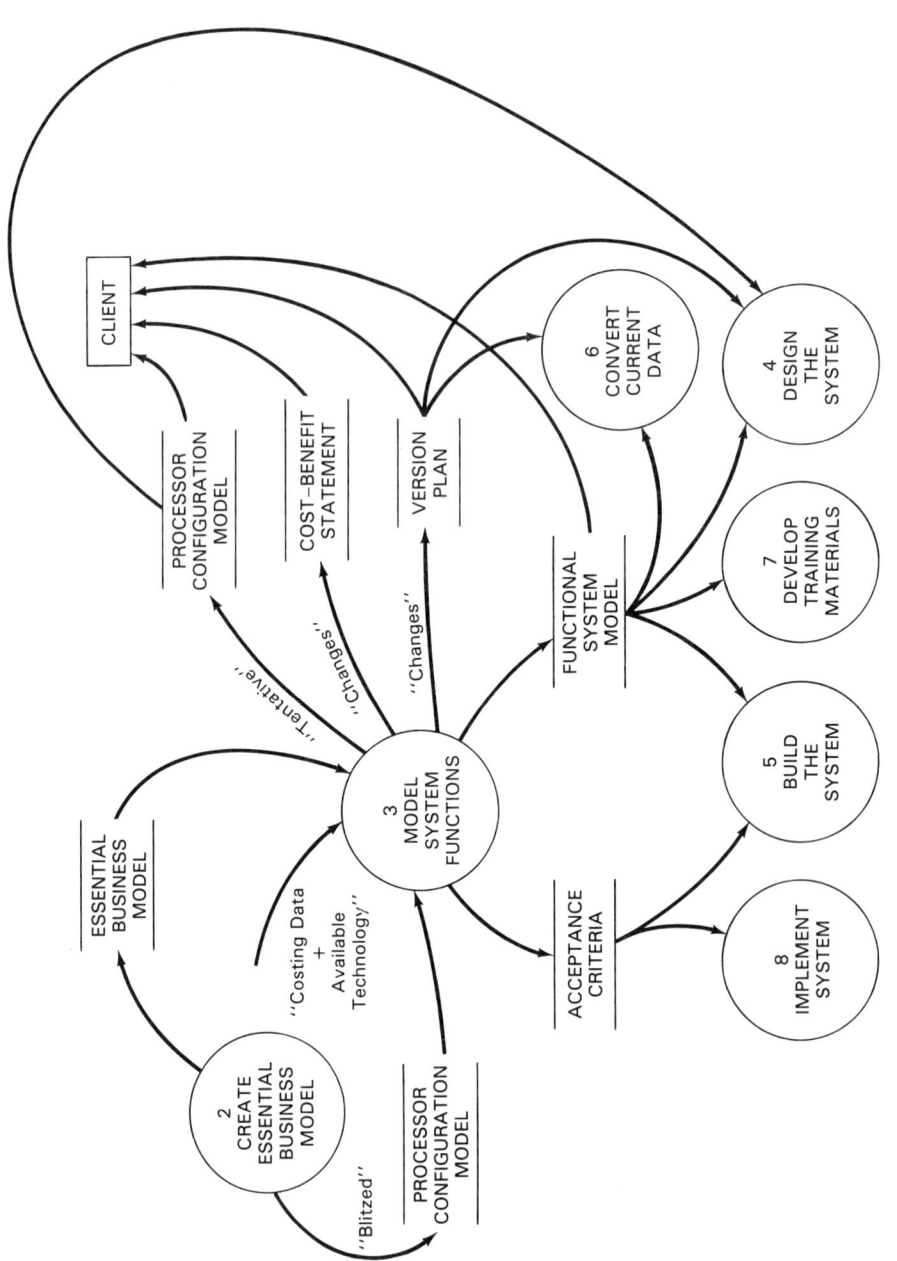

Figure 12: Neighborhood Diagram—Model System Functions

concept and borrowing the processes defined in this book. If, however, the client chooses to fix a few of the things wrong with the current system, the details of the *Model System Functions* process may be significantly different at lower levels. (It is for exactly this reason that a life cycle modeling strategy is important: at any stage of the project, the client may make decisions that radically alter the course of the rest of the project. At such a time, the model of the process of systems development may also be altered to reflect the change in direction of the project.)

Second, the Essential Process and Information models must now be expanded, such that when the Functional System Model is completed, it is totally predictive. What this means is that we must be able to walk through the system's reaction to any possible stimuli, and observe from the information in the model that the system makes the correct responses and that all data needed by the users are available. In short, the Essential Business Model describes only that information and those processes that must exist to carry on business in an ideal world; the Functional System Model, on the other hand, describes all data and processes required by the client and users in the real world, though without stating specifically how functionality and the database will be implemented, nor how processors will be applied to the various functions of the system. Both the client and the project manager must come to the realization that to deliver a successful systems model, the Functional System Model is complete not when money runs out but when it is totally predictive.

Once this (almost) technologically neutral model of the system is found to be acceptable to users, the "automation boundary" is selected by the client, using the preliminary Processor Configuration Model agreed on when the Essential Business Model was created previously. This is done with a little bit of help from some technical wizards who advise, *but do not decide*. The Cost-Benefit Statement is then revised on the basis of what's to be automated and what's not and what processors are to be used, and so is a delivery schedule, or Version Plan. Because automation boundaries, Version Plans, and Cost-Benefit Statements can change dramatically, depending on what is automated or which Version Plan is selected, this process is highly iterative. If there ever was a time when a Development Team could go through many, many what-if scenarios with a client, this is it!

The updated Version Plan can now be used to do some of the analysis related to conversion of data, so that the conversion team will clearly know what the scope of their work will be. It's also used by the designers to focus on the version of the planned system that they will be working on next.

Also, the Functional System Model is used by the conversion team to isolate the domain of the conversion; by the designers so they know what it is they are designing; by the people who develop staff training materials, so they will know exactly what it is they are planning on training people in; and by the systems builders, so they can prepare user documentation and a **Disaster Recovery Plan**.

The only completely new deliverable that comes out of the function modeling process is the list of Acceptance Criteria. The Acceptance Criteria reflects the functions contained in the process model and the policy of the information model. In a sense, it's a checklist that details what the system must be able to do before it will be acceptable to the client. It will also reflect requirements imposed on the client and Development Team by external sources, such as interfaces with other systems, database administration, the DP operations department, and regulatory and legal compliance requirements.

The Acceptance Criteria document is particularly valuable to people outside the project who need a detailed checklist of what the system does (such as audit and security or comptrollers). It also serves as a good initial working document for the **Testing Team**.

3.4.4 Design the System (Fig. 13)

Purpose: To analyze and prepare a model of how the system will carry out the specifications of the Functional System Model.

Input:
* Functional System Model
  > Process Model
  > Information Model
* "Tentative" Processor Configuration Model
* Version Plan
* Acceptance Criteria

Output:
* Design Blueprint
  > **Dialogue Design**
  > **Structure Charts**
  > **State Transition Diagrams**
  > **Data Dictionary**
  > **Module Specs**
* **Database Design**
* **Test Plan**
* **Processor Configuration Model**
* **System Support Requirements**
* (Revised) Acceptance Criteria

The Functional System Model has provided the client with a very detailed picture of the "what" of the system. The task of the design segment is to pin down completely the "how."

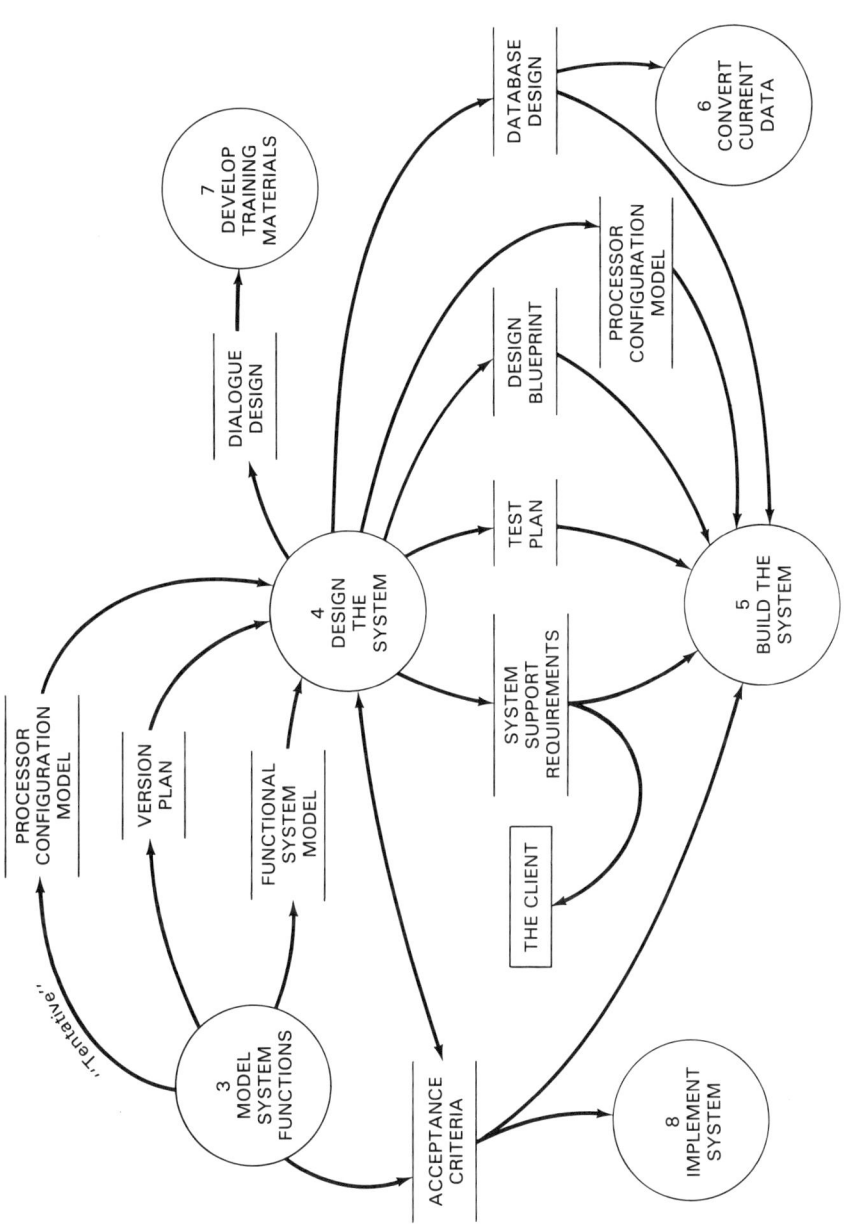

Figure 13: Neighborhood Diagram—Design the System

Now is the time when all physical elements of the system must be completely specified: dialogues, screen hierarchies and controls, databases, hardware, forms, system interfaces, and the like. System Support Requirements are needed by the client so that office equipment, forms, and staffing needs can be arranged. The Design Blueprint, with its components, is needed by the construction team, so they'll know exactly what it is they are building. They will also test the new system against the Test Plan. Finalizing the Processor Configuration Model is necessary; using it, the builders will be able to prepare an **Operations Flowchart**. And because the database has now been designed, it becomes possible to complete the entity analysis, normalizing, and conversion of existing data. Earlier speculation about conversion, although it set boundaries, was really just educated guesswork.

After all these "physical" components of the system have been finalized, the Acceptance Criteria can be upgraded.

Because all functions have been finally allocated to agreed-upon processors (including people) and their implementation mapped out, it's now possible to produce a credible fixed price quotation for the remainder of the project.

Design is often the most time-consuming of the nine high-level processes shown in the level 0 diagram of the development process (Fig. 8).

3.4.5 Build and Test the System (Fig. 14)

Purpose: To deliver a fully tested version of the system to the client.

Input:	*	Design Blueprint
		> Structure Charts
		> State Transition Diagrams
		> Dialogue Design
		> Data Dictionary
		> Module Specs
	*	Database Design
	*	Test Plan
	*	System Support Requirements
	*	Acceptance Criteria
	*	Functional System Model
	*	Processor Configuration Model
Output:	*	**System Release**
	*	Operations Support Procedures
	*	User Documentation

The primary purpose of this process is to deliver a working, fully tested system to the client.

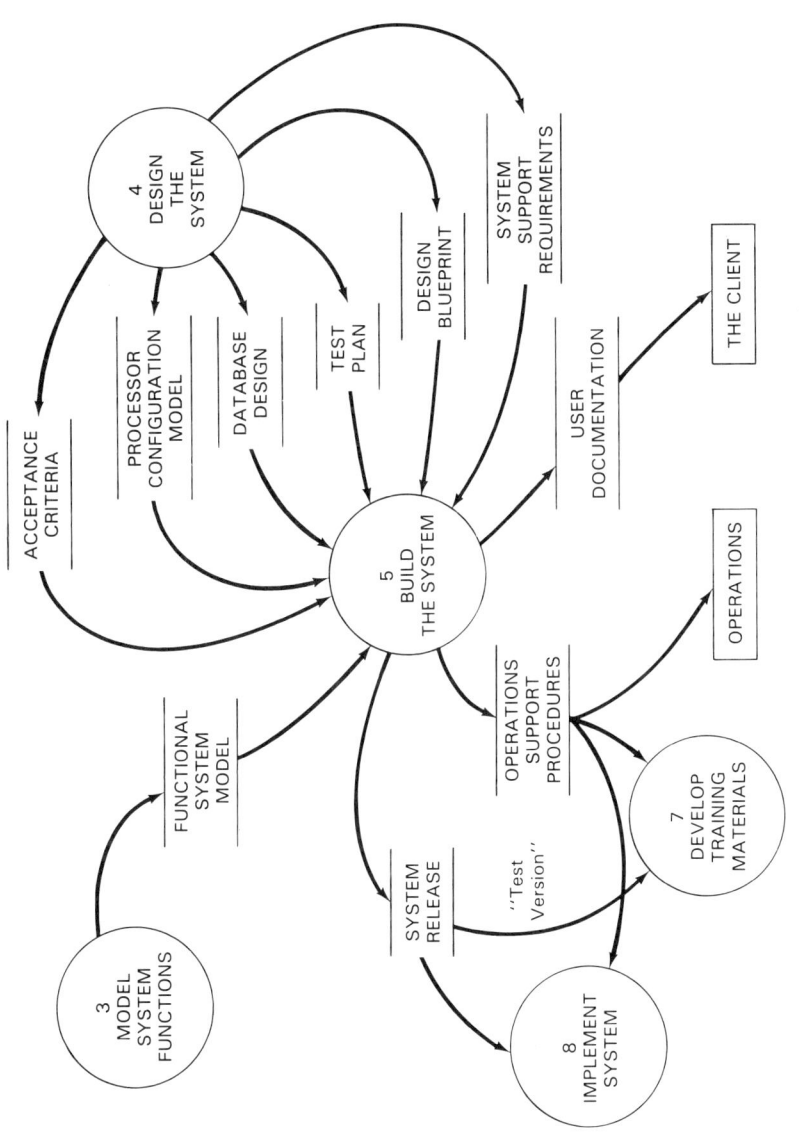

Figure 14: Neighborhood Diagram—Build the System

The System Release includes all software and automated procedures created as part of the project. The detailed descriptions of the manual procedures belonging to the client are contained in User Documentation. The Operations Support Procedures contain all those things Operations needs to know in order to run the system and recover from any problems, such as data entry errors, application program problems, systems software problems, hardware glitches, and acts of God. A "test version" of the System Release is needed to prepare training materials for users of the system.

In a completely manual system there may be very little to do during the "construction" phase. If our staff are our processors, the "programming" of our processors is described in the "design" and carried out in the "training" process. However, there may be other processors involved that support the manual side of the system that do require "construction"; installation of telephone systems, conveyor belts, and any alterations to the office environment can be considered part of the "construction" process.

Needless to say, the process of "building the system" is, to many people, where the real action is. Of course, without all the stuff that went on before, such as information and process modeling, there wouldn't have been any action at all (although fans of prototyping would probably debate that!). Also, if "versioning" is a chosen strategy, then there may be all kinds of results out of this process before all the functions of the system have been completely specified. As Tom DeMarco once said, "You don't need to know much about the doorknobs to pour the foundation!"[7]

3.4.6 Conversion (Fig. 15)

Purpose: To prepare any existing data for live operation under the control of the new system.

Input:	* Database Design
	* (Current) Database Design
	* Functional System Model
	* Version Plan
	* "Existing data"
Output:	* **Converted Database**

This process gets the existing data prepared for live operation under the control of the new system.

The complexity of the conversion process will vary widely from one project to the next, depending on the extent of the change needed. If the change is from a manual system to an automated one, the conversion

---

[7] This quotation actually came from an article in the *New York Times* in August 1979, in an article about "fast-tracking" the construction industry, but DeMarco is the one who first told us about it.

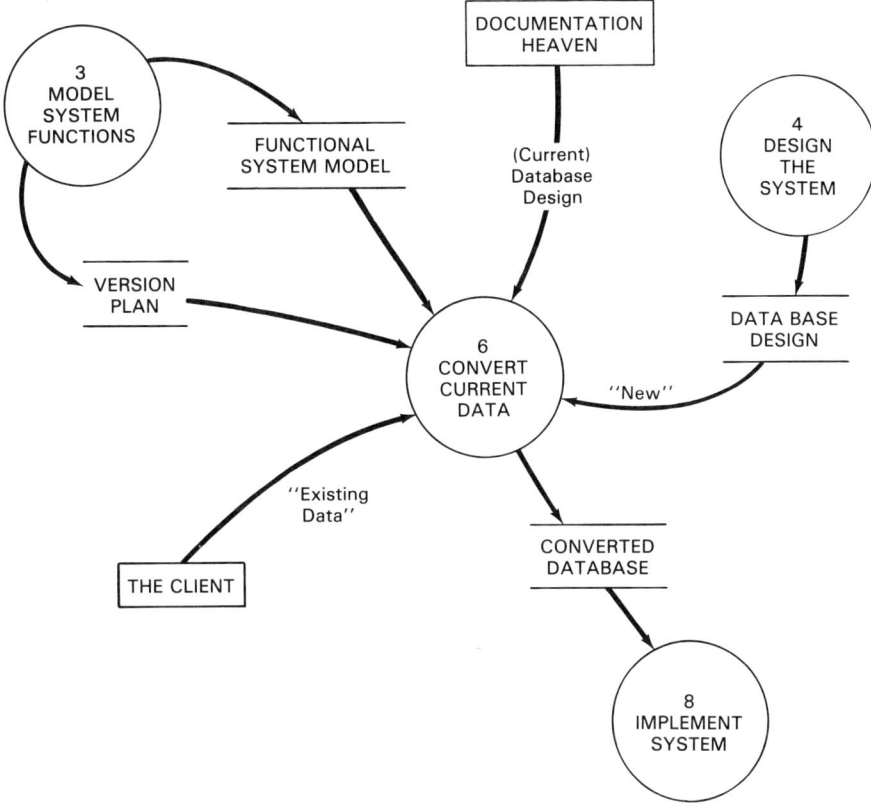

Figure 15: Neighborhood Diagram—Convert Current Data

processs may be merely masses of data entry. There could also be tremendous difficulty in capturing the data, depending on how the data were kept—or weren't kept—which could become a real problem when history is needed in a new system.

An excellent example of massive data conversion related to the implementation of a new system is described in *Up and Running*.[8] H. Dines Hansen, who led the project, describes in detail the conversion of all of Denmark's paper bond certificates to an automated accounting and registration system. Needless to say, the 750 person-year development project went through a very complex mass data conversion.

Going from an automated to a manual system (though we've never seen this done) would presumably require the formatted printing of all existing data.

To make sure only the right stuff (and quantity) is converted, the conversion team will need the Version Plan and the Functional System Model. Only after real data is applied to the Database Design can the

[8] H. Dines Hansen, *Up and Running* (New York: Yourdon Press, 1984).

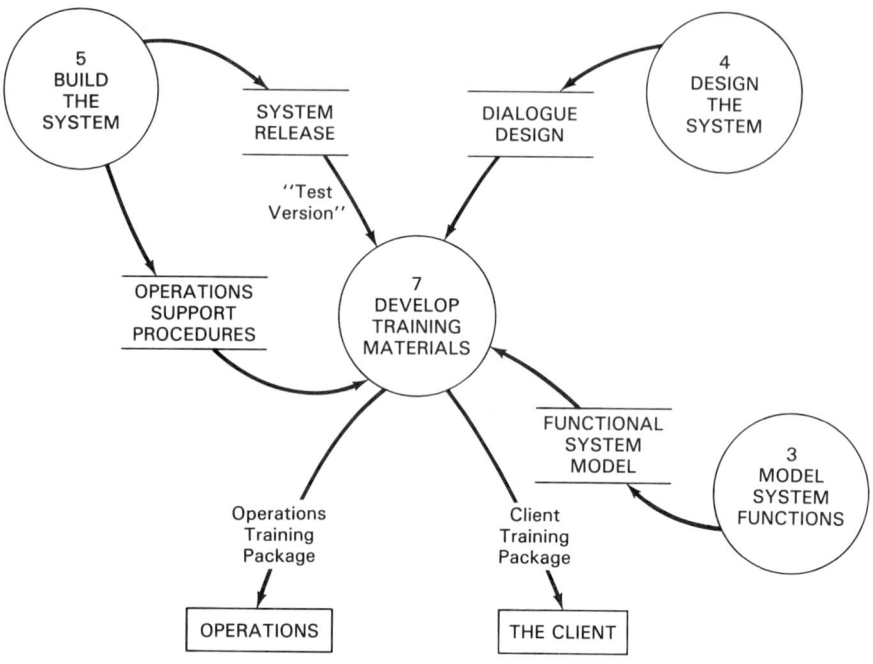

Figure 16: Neighborhood Diagram—Develop Training Materials

Development Team actually implement the system. The difficulty of data transformation from one format to another will be a function of the processors and access methods involved.

3.4.7 Develop Training Materials (Fig. 16)

Purpose: To prepare the necessary training materials so that the system can correctly be made available to users, and users will know how to get the system to respond to their needs.

Input:
* Functional System Model
* Operations Support Procedures
* Dialogue Design
* System Release (test version)

Output:
* Client Training Package
* Operations Training Package

In this process, all the materials needed to train the client and user community, and the operations team, are produced. As well, the process of delivering the training may be seen as either part of this process or as a client responsibility. Whichever it is, it must be clearly understood by all involved, and modeled in the project life cycle. In our example, we have chosen to leave the actual training process to the clients.

Ideally, training should start long before implementation. A good "structured project"—with users (including clerical level users) involved on a daily basis throughout the development process, and actually doing some of the systems modeling—is ideal for this because they acquire the skills gradually over the life of the project rather than in a single, massive effort.

Training is perhaps the single most critical activity for all parties, since the success of training users will ultimately determine the acceptability of the system and its components. Users must not only know how to use the system but they must believe in it, understand it, feel comfortable with it, and recognize it to be an enhancement of their own position.

Without the willing acceptance and enthusiastic participation of users, no system, no matter how good it is, will ever reach the level of productivity that its engineers planned for it.

One of the greatest benefits of having users on the Development Team is the fact that the education of those users is ongoing throughout all the development processes in which they are involved. Not only will they learn about their needs as users, and translate those needs into systems models (since the specification process is very much an iterative, refinement type of activity even for the users), they will also learn an awful lot about the process of developing systems (good for future projects). The repeated modeling and walkthrough sessions will develop users who not only have a thorough understanding of their new system—before it's implemented—but a deep commitment to its success.

The most successful training is working closely with users as partners in development.

3.4.8 Implement the System (Fig. 17)

Purpose: To formally release the system.

Input:
* System Release
* Converted Database
* Operations Support Procedures
* Acceptance Criteria

Output:
* **System in Production**
* "Required changes"

Implementing the system involves the formal release or installation of the system and any special post-implementation monitoring of the system that may be necessary.

Many of the activities of implementation are administrative by nature, such as moving documents to libraries and programs and procedures into

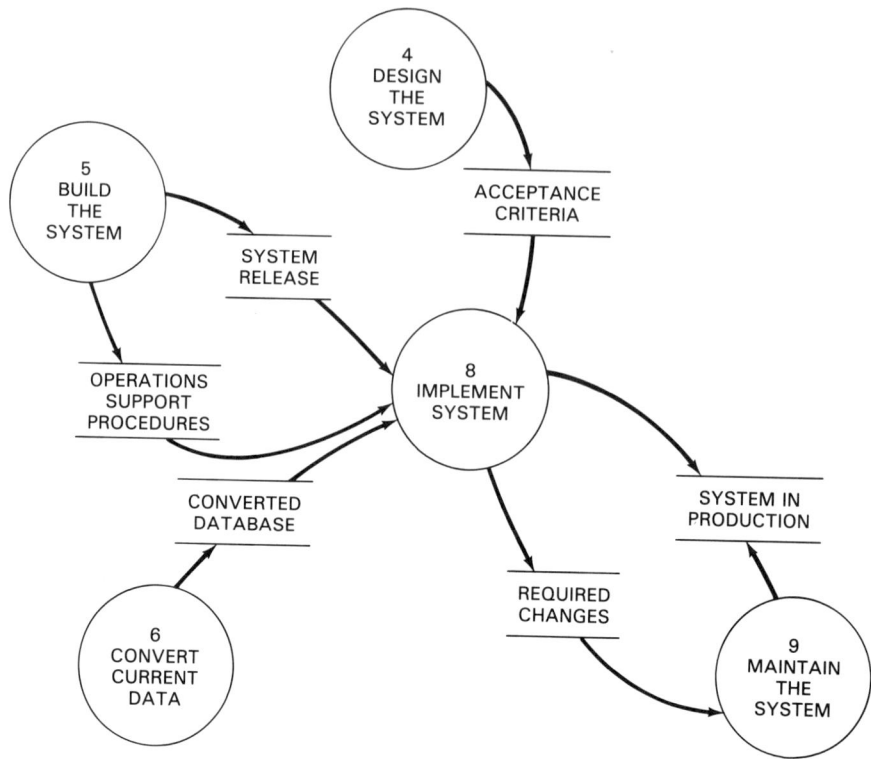

Figure 17: Neighborhood Diagram—Implement System

formal production environments. The model of the implementation process developed for any given system release will have to reflect the specific requirements of that release, especially if there is a period of parallel processing or if the data is very critical or sensitive.

With the completion of this process, the development phase of the project's life is over. A system has now been born!

One more activity should take place at this time, however, And that is the postproject review. This should take place in order to assess the reasons for problems *and* successes within the development process just completed. It is only through the study of the projects we undertake that we can improve what we do when building systems in the future, as well as improving other related skills (such as cost estimating).

3.4.9 Maintain the System (Fig. 18)

Purpose: To monitor and maintain the system, ensuring its availability and quality.

Sec. 3.4   The SuperSet Model's Process Abstracts   55

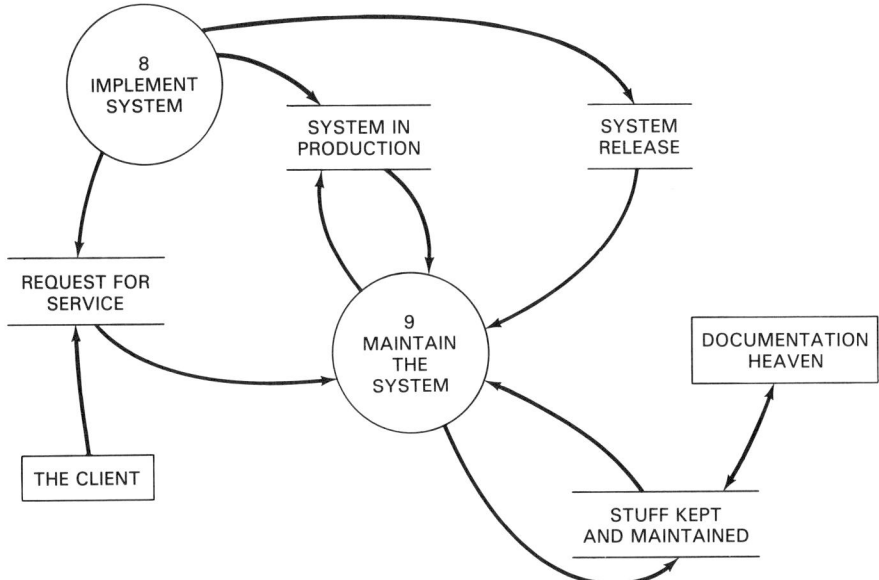

Figure 18:   Neighborhood Diagram—Maintain the System

Input: *   System Release
            *   System in Production
            *   Request for Service
            *   Stuff Kept and Maintained

Output: *   System in Production (revised)
            *   Stuff Kept and Maintained (revised)

It may come as a bit of a surprise to see another activity in the model after implementation! Surely, after the Champagne has been cracked, and weekend passes have been reintroduced for the Development Team, isn't the project over?

Not quite so fast, my friend!

Certainly the bulk of concentrated activity is over. But the system isn't dead; it has just been born! As a result, an extended program of preventive medicine is necessary to keep it in good health.

Changes to the system can originate in a variety of quarters, and for a variety of reasons:

- The client can ask for changes due to
  1. changes in business requirements (either as a reflection of adapting to the marketplace or of externally imposed requirements, such as government regulations);

2. a need for more functional support;
  3. detection of errors in the system.

- Data Processing can initiate changes due to revisions to operating systems software, teleprocessing changes, DBMS changes, and a variety of other technical changes that are independent of the client's business.
- Other external or corporate agencies (for example, other departments that may or may not be a "terminator" to the system currently) may request changes in compliance with new policy or legislation.
- The people who are responsible for quality management may request modifications to accommodate changes in the interfaces resulting from any of the above causes.

User-initiated enhancements are really small complete projects in their own right. The SuperSet System Development Model can be used to select those activities that will be needed to implement the requested changes. The elements of the SuperSet needed to model a change will depend very much on the scope of the change.

If there is a change in major business functions, then the project life cycle will be fairly complete at implementation time, because all major models produced will have to change, including the Essential Business Model.

If a function is being added to the system that falls within the scope of the *current* Essential Business Model, then only the Functional System Model, the Design Blueprint, and the code must change.

Similarly, if an *existing* function (which is currently handled manually) is to be automated, only the Design Blueprint and code have to be updated.

Fixes due to simple programming errors will only require a change to the code. But if the defective code is due to an error in creating one of the models earlier in the development process, then *all models from the source of error on will have to be altered*. Badly done business analysis could affect the Process and Information Models, which would cascade into further problems in the Design Blueprint, which, in turn, would possibly have an even greater effect on the Database Design and the code. If the Essential Business Model is simply wrong, you might as well start again. But if it is "too physical," it won't cause an "error" in implementation, but it may constrain the physical system and make it less flexible, less "perfect," and therefore more difficult to enhance in the future. It pays off, then, to get your best people concentrating their efforts on the front end of systems development so that the back end (the implemented system) will be relatively easy to maintain and enhance.

Within a maintenance project, the use of the SuperSet System Development Model will ensure that all the changes that must be made *are made*, both to the implemented system and to the documentation (Stuff Kept and Maintained).

3.5 REFLECTIONS ON THE OVERVIEW CONTEXT

The eight major processes of the level O diagram of the SuperSet System Development Model (Fig. 8) reflect the majority of the activities that are necessary to carry through the implementation of a new system. The ninth process is included so that systems can be well taken care of and made to be better. This permits an organization to have systems rather than projects, and thus completes the systems life cycle.

These nine essential processes are necessary whether the system to be developed is all manual, fully automated, or somewhere in between.

Not all projects will require all activities, however. Some projects may need activities that aren't shown here. What must be realized is that every development or maintenance project is unique and will require a unique development process.

We should also reiterate that we created this model of the development and maintenance process using structured analysis techniques. Your own model of the development process should be built using the tools *you* will be using for analysis and design.

The life cycle shown here may not work for all projects, nor in all environments. But every project must have a defined life cycle and its processes understood by all those involved: client, users, and Development Team. Through the use of a superset approach, the client and Development Team managers can agree on which tasks are necessary to reach their specific goal. Since every task should be essential, and every participant should understand why, the project life cycle model becomes a project management tool.

The life cycle model produced by our superset game plan is, therefore, both a management *and* a project control tool. It won't manage the project or team at the task level, but it works well on a broader scale.

Remember also that the activities selected for any given project are not serial, but asynchronous. The major processes specified for a project are not synonymous with project phases. In fact, as we mentioned before, reality is not serially phased, which is why serially phased projects have so many problems. Many different activities may be going on at the same time; specific activities may be repeated over and over again, "until it's right"!

There is also no specific relationship between the number of original stimuli and the number of implementations. One major request, which will either revise or create an Essential Business Model, will, with a Version Plan, imply several iterations of analysis, design, construction, and implementation of system releases.

Every company will have its very own business and technological—not to mention administrative—environment that will allow it to tailor a superset of activities much more accurately targeted to its own environment than we could show in a book of reasonable size. In fact, we have tried to model a superset that is applicable to most commercial systems. It's our hope, though, that people will examine their situation and say, "What is the superset of activities for my company?" and then set up that superset and use it for modeling specific projects. Then, with the hindsight allowed by postproject reviews, the superset can be further refined after each project, making it a more valuable and powerful tool as successive systems get built.

CHAPTER 4

THE SUBSTANCE OF THE MODEL

What is the question to which this is the answer?

E. F. Lloyd Hiscock

4.1 INTRODUCTION

In light of the differences between projects, it's difficult enough at a high level to identify a superset of processes that will be common to all projects. At the detailed level, it's virtually impossible. At the lower level the impact of the development environment becomes much more pronounced, and in some cases it's not possible to put together into one model of a lower level process the activities needed in two radically different kinds of projects.

We will present models for all of the nine major activities, but it's much more important to realize at this level that these are hints and suggestions; they may serve as a starting document, but everyone must create his or her own lower level supersets to reflect individual reality.

Once those lower level models have been set up in each organization, the cardinal rule of the *game plan* applies:

ADAPT, DON'T ADOPT!

Each team's project manager must choose only the essential activities from each set for its life cycle. Each activity in the superset must be approached and asked:

- Is the activity necessary?
- Is this the most cost-effective way to move toward project completion?

The superset approach mentions a lot of activities that require some very specific skills (such as analysis, process modeling, information modeling, systems design, dialogue design, program design). The assumption is made that the members of the Development Team will have the necessary selection of those skills to do the work. The SuperSet System Development Model—and this book—may show *what* is to be done, but it doesn't show *how* (although we'll hint a little bit). Major works on the *hows* have already been done by others. Each company, each development group, must have a *toolkit* of skills available and recognize that the development of these skills is a necessary cost and an investment in the building of a strong systems development team.

In order to achieve the most successful use of a superset approach to systems development, the use of walkthroughs at all stages of systems development is mandatory. In *Structured Walkthroughs* Ed Yourdon examines this most successful method of peer group review of functional specifications, design, and code.[1] The text explains how to use walkthroughs to detect errors in the code, inefficiencies in design, and analysis errors.

Holding several walkthroughs for all major deliverables may increase front-end costs a little, but the back-end payoff is very substantial. Walkthroughs are essential because they ensure that quality and user commitment to the project are built in at every step along the way. Each function and process will be seen to be necessary and correct. The savings realized by delivering the right product far outweigh the cost of development walkthroughs. After all, wasn't the original impetus to the use of structured techniques the fact that systems are much cheaper to alter early in the development!

4.2 THE DETAILED DIAGRAMS

Detailed lower level diagrams follow for each of the nine major activities. The processes in each diagram will be discussed briefly. In a more formal model, this discussion would be represented by process specifications, but what is important here is that the general flow of activity (and transformation) is understood. Remember that the superset you are seeing here is a generic example. Before you use a superset approach on a specific project, you will have to build your own superset model.

Not all data flows will be explained. We suggest you keep your finger in the Dictionary (Chapter 9) as you read through this section.

[1] Edward Yourdon, *Structured Walkthroughs*, 3rd ed. (New York: Yourdon Press, 1985).

4.2.1 Set Up Preliminary Agreement (Fig. 19)

The Preliminary Agreement is a small package of three deliverables:

- Objectives & Constraints
- the **Client List**, and
- the **Approval Group List**

It should be possible to set up the Preliminary Agreement quickly. Usually, two days will be sufficient. If it takes more than five days, then too much is being attempted, or stultifying bureaucracy is interfering with the process.

The Objectives & Constraints should briefly summarize the scope of the project, any known external or corporate constraints, and what the client would like to see in terms of schedules, priorities, and budget. It should also specify what the client and Development Team have agreed to do next

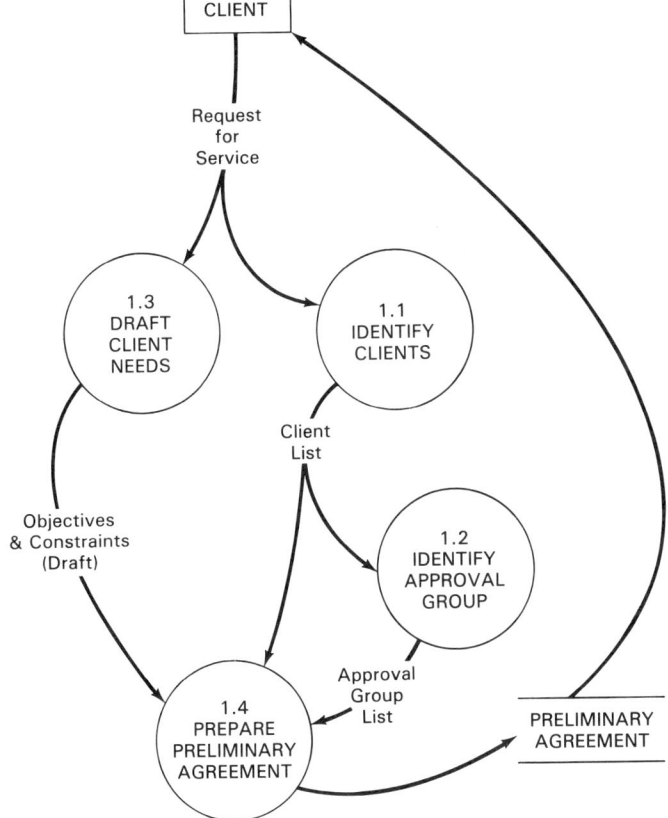

Figure 19: Set Up Preliminary Agreement (1)

(usually an Essential Business Model) and a target cost and schedule for that deliverable.

Although the Preliminary Agreement should be prepared by the client, it will in fact usually be written by the person on the team doing the work. This may be a procedures specialist, somebody from the data processing department, or an external consultant, but it is important that the person write from the viewpoint of the client. Involving the client directly in preparing the Objectives & Constraints is vitally important; it could set the tone for the remainder of the project.

The Client List will identify main people from the client's immediate area who will be involved in the development process and in reviewing the accuracy of project deliverables.

The Approval Group List identifies the people who will be formally responsible for "signing off" (or whatever it is you do in your company) the major deliverables. The approval group is usually client and user management, and perhaps management from other areas of the company. It may also include auditors, database administrators, and others whose presence may be suggested (or mandated) by corporate policy.

In short, the Preliminary Agreement should be seen as a contract between the client and the Development Team. In fact, the Preliminary Agreement should be much like the sort of contract you would draw up if you were engaging a consulting firm on a time and materials basis. It's just an agreement to start work, specify who does what, for whom, what the major deliverables are, what the expectations are, and who gets to make approvals. With this basic information as a guideline, the Development Team can start to develop the Essential Business Model.

4.2.2 Create the Essential Business Model

While it's fine for a company to establish formal education and apprenticeship programs for analysts and designers, a little bit more is needed. A commitment to a basic philosophy of *Do Good Work* is needed. And "doing good work" entails understanding the many components that go into developing quality systems.

Almost all of these elements are suggested or contained in this book, either in the main body, in the Dictionary, or in the Manager's Notebook. But another view won't do any harm.

Front-end Business Analysis

Systems development must be more aligned with fundamental business objectives, not just administrative support procedures. Systems developers and planners need a methodology that can respond to essential business events. Without such a methodology the systems developer's mind-set remains current, and there is little opportunity to discover how things should be.

Sec. 4.2 The Detailed Diagrams

Business analysis transcends automation boundaries and can be done through the development of Essential Process models and Essential Information models. Combined, these are called the Essential Business Model. EBMs can be developed at any desired level; that is, there can be an EBM at the application system level, at the departmental level, at the business unit level, or at the corporate level. The fundamental difference between an EBM and a BSP (*Business Systems Plan*)[2] is that an Essential Business Model can actually be implemented.

The delivery of a *current* Essential Business Model, although new versions rarely change substantially, permits you to evaluate whether or not the current business architecture can be improved by changing fundamental data relationships, departmental responsibilities, or administrative processes.

The availability of such a business architecture, in the form of a modifiable *model*, also affords the opportunity to play what-if games at a departmental or corporate level. Examples would be the assessment of impact if your company introduces a specific product to the marketplace. Is the *essential business system* prepared to handle the processing of the new business? Could you quickly develop the *physical systems* to support the new business? When planners, architects, and marketing staff have an Essential Business Model, they are able to analyze the effect of business decisions and technology.

Especially in mature companies, most of the work involved in creating an Essential Business Model is work that shouldn't have to be done, except in very rare cases. But because of a lack of rigor to the deliverables of the development process in the past, Essential Business Models *do* have to be created in almost every case of systems development work today. (We'll look at how to avoid this problem when we discuss process 9, *Maintain the System*.) Most of the real work in this process is inside activity 2.1, *Create Models of Essence* (see Fig. 20).

Most software repair is a response to a symptom, that is, poor code and design. The cause is insufficient detail and quality of analysis; that is, the Essential Business Model and Functional System Model either don't get created or they are superficial. There is ample proof of this in the software industry, an industry that generally puts more time and effort into building software than businesses do with custom-built systems. The following is a direct extract from a typical software license agreement:

> This program is provided "as is" without warranty of any kind either expressed or implied. The entire risk as to the quality and performance of the

---

[2] BSP is a methodology out of IBM which addresses and identifies data classes, processes, and their relationships. Michel Veys of IBM in Brussels, however, has now developed a similar automated methodology called ISS, which he suggests will make the traditional BSP obsolete. So much for BSP!

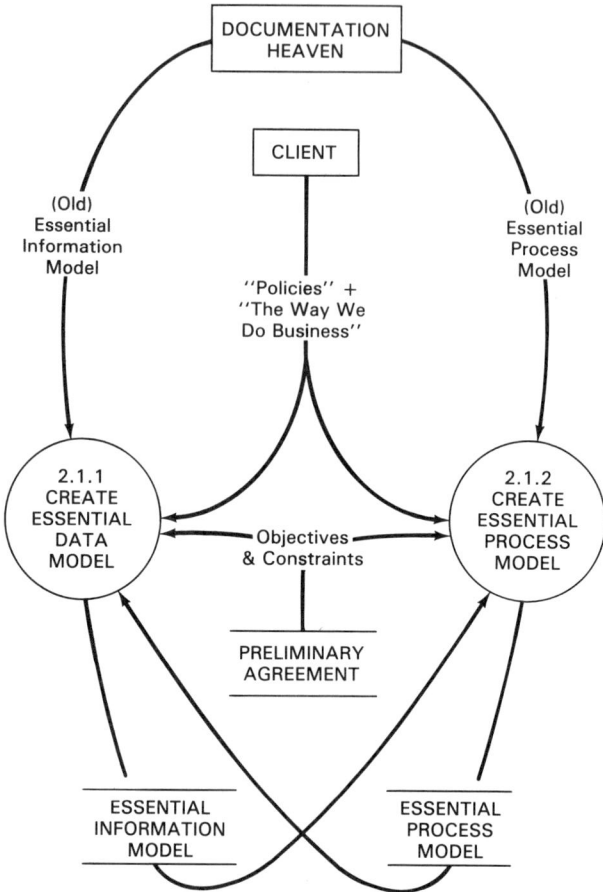

Figure 20: Create Models of Essence (2.1)

program is with you. Should the program prove defective, you assume the entire cost of all necessary service, repair or correction.

When insufficient modeling has been done, the systems that result will often fail to properly isolate business functions from physical considerations and will not be robust. Rather, they will be difficult to adapt to either new technology or changes in business.

To improve quality and productivity, we strongly recommend you put the resources and effort in the front end of the project, appropriately staffed with user personnel. That is, use people who understand the "essential" functions and data, or *innermost substance*, of the business for which the system is proposed. Use people who understand that each system has a part that resides *inside* an automation boundary while another part is *outside* that

Sec. 4.2   The Detailed Diagrams   65

boundary. Staff the project with a Development Team that understands they have to comprehend and define the *what* before they figure out the *how*.

Creating the two component models of the Essential Business Model and the subsequent Functional System Model is the most complex, critical, and time-consuming part of the entire development process. If the developers haven't got the essentials right, the rest of it will be even less right. Once it is right, however, programming and implementation are relatively inexpensive and fast.

Before looking at the first significant process in the systems development life cycle, however, let's discuss the people who make it all happen.

Throughout this book we refer to the *Development Team*. Teams of one don't usually work very well, so we've made the assumption that every new development project or substantial maintenance project should consist of at least two people, and preferably more. This is the group of people associated with analysis, design, and quality engineering of a user's system. We also feel that the team should include what has loosely been termed "user analysts."

Team members are not necessarily all full-time members on a project; team members can assist as their particular skills are required. Also, team members can often be different people from one phase of development to another. That is, analysis teams, which are heavily user-based, can be staffed differently than design teams, which are oriented more toward technical skills.

An example of a design team could be:

Project Team
    Project Manager
    Design Architect
    Design Apprentice
    Design Apprentice

Consultants
    Project Architect
    Strategic Planning
    Systems Security
    Audit and Controls Specialist
    Database Specialist
    Documentation Specialist

The *project team* are full-time members, while the *consultants* (usually from inside the company) provide expertise as needed. Also, individuals are listed by their role rather than by title. Ignoring titles and concentrating on roles usually sidesteps egos and enhances team cohesion.

The team concept is based on the premise that any one or two individuals cannot possibly know everything there is to know about developing a system. Also, it takes advantage of the synergistic effect a team can provide.

For teams to be successful, however, the members must have a number of common attributes:[3]

- ability to be analytical—an appetite for tearing systems apart, plus a merciless desire to scrutinize each piece;
- ability to resist doing something because it's "convenient";
- commitment and ability to communicate clearly with other humans, not just machines;
- ability to tolerate uncertainty, to take satisfaction in identifying questions as well as answers;
- ability to withstand the abuse that always accompanies any effort to "do good work";
- ability to distinguish fundamental business policy from issues of implementation, and then, during design, to implement the business policy in the most flexible and professional manner possible;
- ability to discern between nice-to-have and essential;
- ability to use the tools of analysis and design instead of being a slave to them, to recognize their limitations as well as their strong points;
- willingness to invent whenever the need arises; and
- at least one member who understands the politics of projects.

The team concept in systems development is almost as important as using some kind of systems development life cycle, and shouldn't be treated too irreverently. To help team members communicate well, in a controlled manner, it would be worthwhile reading Rob Thomsett's *People and Project Management*.[4] In this easily read book he describes how to develop successful project management methods that facilitate the flow of information among analysts, designers, implementors, maintainers, and users while taking into account the needs of these individuals. He also discusses individual motivation, job satisfaction, performance, work attitudes, leadership roles, management issues, and how new software design methods affect control of systems projects. An interesting supplement to Thomsett's work is Philip C. Semprevivo's *Teams in Information Systems Development*.[5] Semprevivo defines a rational methodology for applying strategies

---

[3] With thanks to Gary Schuldt, a former Yourdon colleague, who defined many of these attributes for data modeling experts; they apply equally well to other systems development professionals.

[4] Rob Thomsett, *People and Project Management* (New York: Yourdon Press, 1980).

[5] Philip C. Semprevivo, *Teams in Information Systems Development* (New York: Yourdon Press, 1980).

Sec. 4.2   The Detailed Diagrams

based on management theory and organizational development theory to the special problems of systems development groups.

Having looked at the people who do the development work, we can now proceed to address the work to be done. In conjunction with our discussion, look at Fig. 21.

You will notice that process 2.1, *Create Models of Essence*, has attached to it a mysterious terminator called Documentation Heaven. Put a dog-ear on this page so it will be easy to find when you want to return to it, and then turn to the Dictionary and look up Documentation Heaven.

Many of us were weaned on the techniques of Tom DeMarco and memorized at an early age the ritual chant of ". . . current physical, current

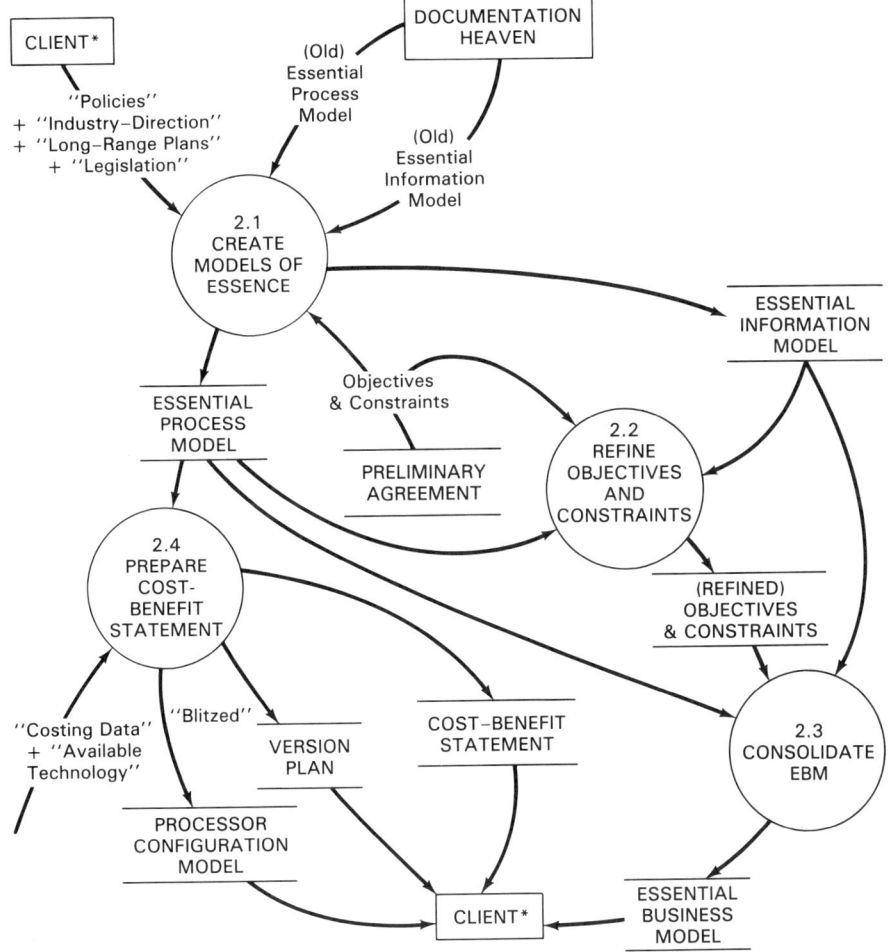

Figure 21:   Create Essential Business Model (2)

logical, new logical, new physical. . . ." Although it wasn't really very clear in the beginning what *logical* meant (was the complement "illogical"?), through osmosis we eventually figured it out, and we thought we were doing what he taught.

But if this were really so, the *current logical* model for a system would now be available in the existing documentation, because the *current logical* we're looking for is simply the *new logical* model from the system's previous implementation.

Documentation Heaven is where those things reside that systems developers are supposed to leave behind. (We also believe in the Easter Bunny!) In reality, such documentation is often completely missing or sadly out of date.

If the current documentation is missing, it will have to be created again. But the client should understand (assuming the documentation *is* taken care of in the future) that after one generation of developing a system, this step of uncovering the *current logical* will not have to be repeated. We will discuss it in more detail when we review the *Maintain the System* process.

But the process of modeling the *current physical system* is also fraught with danger. This is when the Development Team could easily get sidetracked and become involved in an archeological dig that could very well take forever. The temptation to model the existing system in great detail can result in almost unattainable goals; Ken Orr called this part of the analysis process "the Bermuda Triangle of systems development."[6] If it is necessary to produce a model of the current system, it should be done as quickly as possible and the result should be as lean as possible. After all, the reason we create *current* physical and logical models—which are irrelevant once the new system has been defined—is to give ourselves temporary steppingstones in the evolution of the new system.

Since this activity shouldn't have to happen anyway, it's politically expedient to convince the client to undertake the modeling of the existing system *independently* of the new project. This activity can often take a couple of months (or even many, many months). If the client confuses the delivery of the Essential Business Model (which is really a very small deliverable) with all the time and effort put into documenting the current system (which the Development Team is going to throw away anyway, almost immediately after it's produced), the client may not feel there was good return on the investment after so many months of waiting.

If "current systems modeling" is separated from the development project, a Development Team consisting of a couple of senior analysts—having access to senior client staff with full knowledge of the business they

---

[6] Ken Orr's *Structured Systems Development* (New York: Yourdon Press, 1977) discusses a method of logical analysis, design, and development that can be used on any kind of system, computerized or not.

are in, and using event-based analysis techniques—can probably deliver a reasonably good Essential Business Model for just about any project of any complexity within a month. Smaller projects could take much less time.

However, with more complex systems, it is sometimes desirable to conduct "scouting blitzes." This will assist in gaining a better understanding of what will be within the scope of the project, and to improve the initial estimates of time and resources needed to deliver an Essential Business Model.

Scouting blitzes should be done by selecting two analysts to reconnoiter the target system a few weeks before the main Development Team starts tackling the problem. This blitz should take no more than three to five days.

The scout team will produce a quick (and not so dirty) Essential Process Model of the user's business. These scouts will also do a very superficial Essential Information Model, roughing out a model of the data used in the essential processes. Through the blitzed versions of these models, basic policies and functionality will be identified, at a very abstract level. It's very important for the scout team to identify, as much as possible, existing and probable interfaces with other departments and systems when doing this reconnaissance.

These blitzed models can then be used for:

- better estimates of project complexity;
- planning, selecting, and organizing the main Development Team;
- judging early work deliverables;
- getting the client to understand that the company and its business is more complex than a single department; and
- getting a more reasonable budget allocation from the client.

A blitzed Essential Business Model can provide valuable reconnaissance data of the project effort ahead. It also gives the project manager a decent interval in which to develop work plans and estimates for the main Development Team, based on the discoveries of the advance team. Blitzed models are not substitutes for full-scale models.

Creating Essential Business Models

The essence of a system consists of all aspects of the system that must exist, regardless of how the system is implemented.

The Essential Business Model is the combination of the Essential Information Model and the Essential Process Model (or *new logical*, as we used to call it). The Essential Information Model describes business subject matter in terms of business entities and relationships, whereas the Essential

70   The Substance of the Model   Chap. 4

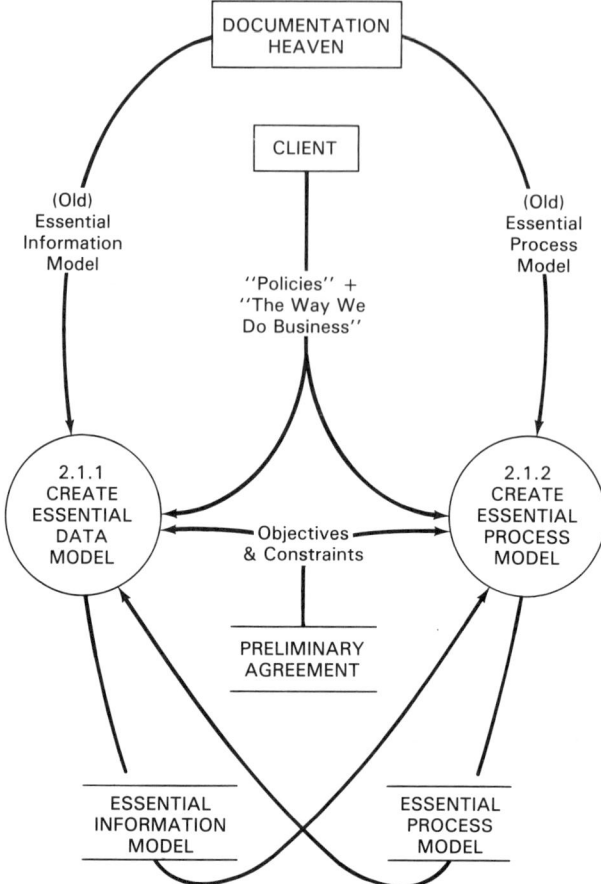

Figure 22: Create Models of Essence (2.1)

Process Model reflects the essential activities and memory of the system. Both of these subjects have been extensively treated by other authors.[7]

Process 2.1, *Create Models of Essence* (see Fig. 22) may appear paradoxical; the Essential Information Model is an input to the creation of the Essential Process Model, and the Essential Process Model is input to the creation of the Essential Information Model. What's happening here is that the two processes are parallel, and that details of each model as it is being created will facilitate the creation of the other.

The creation of correct essential models is vital to the success of the

---

[7] Matt Flavin's *Fundamental Concepts of Information Modeling* (New York: Yourdon Press, 1981) describes basic data modeling as an analytical tool, while Stephen M. McMenamin and John F. Palmer discuss essential models, both the process and data kind, in *Essential Systems Analysis* (New York: Yourdon Press, 1984).

development project. Any errors (or attempts to skimp) at this point will echo with magnifying cost through the rest of the project.

One way of being closer to the mark when creating the Essential Process Model is to start off with an *event list* and then translate the event list into the necessary processes. Events are things that usually happen *outside* the system, or *time-related things* (temporal events) that occur *inside* the system, to which the system must issue a planned response. Events are conditions that occur or a state that is entered that provide some form of stimulus to the system and call for some kind of planned response. Systems do not do things for their own sake; they exist only to respond to specific business *events*.

In order to determine the essential processes to be performed, the Development Team can begin by compiling a list of temporal and external events to the system. A few examples of system events (say, affecting a personnel system) could be as follows:

The Employee

- is hired/terminated
- starts/restarts work
- stops work due to retirement, death, termination
- stays away from work temporarily due to vacation, illness, leave of absence, maternity leave
- chooses benefits including mortgage, life insurance, membership in the credit union, staff association, fitness club, car and/or house insurance, parking space
- is promoted/demoted
- wishes to be paid, and so on.

Each of these events would require a response from the system, either altering the stored information or providing some kind of output from the system.

Each event also provides the *stimulus* to the system, either by providing data upon which the system must act or in the form of a time trigger (daily, weekly, monthly, for example, such as when an employee is to be paid).

To produce the Essential Process Model, the Development Team examines each event and identifies the stimulus (and the source of the stimulus) and then describes the process that will produce the desired response. For example:

Event: Employee is hired
Stimulus: Notification of hiring from Recruiting Department
Response: Record name and address, employee number, Social Security number, position, and other pertinent data

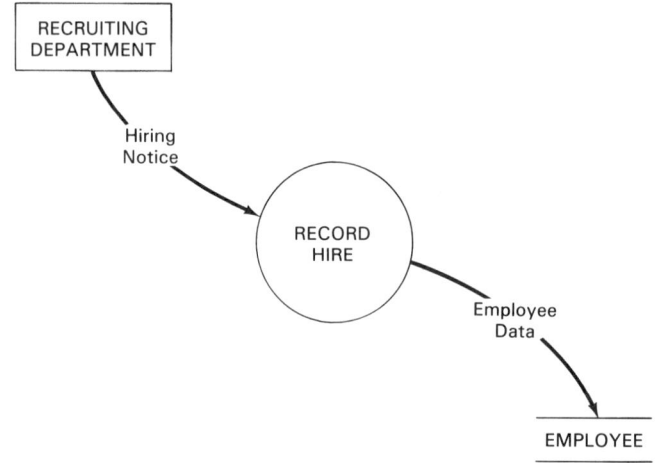

Figure 23: Record Hire

Graphically it could look like Fig. 23.

Process: For each Hiring Notice, record the employee data.

The event list and model are excellent working tools to arrive at a logical model, build the initial data dictionary, and write a specification for the process that addresses the substance of the business needs. It helps to make the process of uncovering essential activities with the user much more manageable. For example, with an event list (developed with and confirmed by the user), it simply becomes a matter of asking the users specific questions related to what must be done to respond to the given event. The user can usually tell the analyst exactly what the stimulus to the essential process is and what the planned response should be. After each event has been reviewed this way, the analyst then would bring all the essential processes carried out by the system into a cohesive process model, properly partitioned and leveled, consisting of data flow diagrams, process specifications, and a data dictionary. The work gets done in manageable chunks, and then gets tied together in a logical, mechanical manner.

Productivity and the Essential Business Model

There are many ways of improving productivity when developing a system, such as using fourth generation tools and prototyping. But such tools generally only affect coding and related activities, which represent a very small part of the entire development process. What's needed is a technique to dramatically speed up the laborious front-end process of *analysis*.

To meet this need, we feel there are few ways that can get you off on a flying start better than event-based analysis and creating an essential model—that is, an accurate model of the innermost substance of the business and the obligatory memory needed to make it work, regardless of

Sec. 4.2 The Detailed Diagrams

the chosen implementation technology. In other words, *do not do a current physical model of the existing system, but jump right into specifying the essential functions wanted* in the new system. And, when you're doing this, *ignore* all edit functions, audit features, approvals, and the sundry other controls that can be integrated later. In our experience, there is no faster way of obtaining productivity gains in the usually long and arduous process of analyzing new requirements.

The preceding example of events in a personnel system is but an introduction to the concept. Since this topic is so important and can gain so much productivity, as well as quality of analysis and time, we would like to go through a much expanded example of how it can work.

Assume you have just been assigned a new project that must be implemented in the next four weeks, and that it has already been turned down by three other project leaders as "work in the impossible region" because of the short time frame. But you're brave. You have just learned about *essential business modeling* and have a background in the structured techniques, so you've decided to tackle the problem. You received the following memo from your boss.

MEMORANDUM
MAXIMONEY, INC.

To: Ted "Rambler" Yento   Date: June 1

From: Verbal Funderbuck
       Vice President, New Business

   Subject: New Currency Exchange System
   Copies: The Big Boss
            Personnel File
            Ted's Team

I was pleased to hear, Ted, that you felt this new system was "no big deal," and that you are putting your career with MaxiMoney on the line with this one. My understanding is that if you deliver what we want, on time, and within budget, you get a couple of weeks off in August; if not, you get permanent relocation to our life insurance subsidiary.

   To review what we discussed at our meeting yesterday:

   MaxiMoney, Inc., has decided to expand its products in the consumer financial services sector. We have acquired the rights to establish a currency exchange booth at Burbank International Airport, in beautiful-downtown Burbank, California. This new business division will be known affectionately as MaxiChange Financial.

   MaxiChange must be fully operational by August 1. Your mission, as agreed, is to define the essential business and memory requirements for the new system, and then get the thing implemented using MagiCode and the new micro relational database product Monolith-R.

MaxiChange will be an independent business division. It will be responsible for paying its own rent, bank charges, personnel, and employee tax withholdings and filing the appropriate employment forms and remittances with government. Galactic Headquarters will look after business taxes and such. Also, it is only interested in receiving balance sheet information on a monthly basis.

Galactic Headquarters will control funds in all banking accounts set up at our MaxiBank subsidiary in the Cayman Islands. That is, if MaxiChange has a funds shortfall at any time, head office Cash Management will provide additional funds as necessary. Conversely head office will remove excess funds, but only at quarter-end.

The business will operate as follows:

- customers can convert foreign currency into local currency
- customers can convert local currency into foreign currency
- customers cannot convert foreign currency directly into another foreign currency
- from time to time MaxiChange must recognize new currencies· this is based on demand and is determined by the independent business division
- currency exchange rates change often (sometimes)· rates will be obtained directly from the Chicago money market using the PC software product $RATEWATCH
- MaxiChange will operate on a fixed rate basis· that is, we charge a flat fee of $5.00 per transaction
- Galactic Headquarters wants profit-loss and balance sheet data on a monthly basis; this is to be uploaded from the local PC to the DataSchwartz 2938 mainframe at head office· head office will design its own reporting mechanism and will be responsible for reconciliation
- MaxiChange has an unlimited line of credit with MaxiBank, which handles all our currency transactions
- MaxiBank charges other divisions prime plus 5 for borrowed money, and pays prime less 5 for excess funds.

We have been blessed with the work of Thusnelda Neusbickle, our founder's niece and our resident Corporate Information Modeling expert. Attached is a high-level view of the static data model and relationships as defined by Thus when we investigated this business opportunity [see Figs. 24 and 25]. It reflects only the perspective of Galactic Headquarters, and not any entities and relationships you may discover internal to MaxiChange. Go by this model. It is Corporate Truth.

Your user is Horacine Clutch. She is cantankerous and mean.

Good luck!

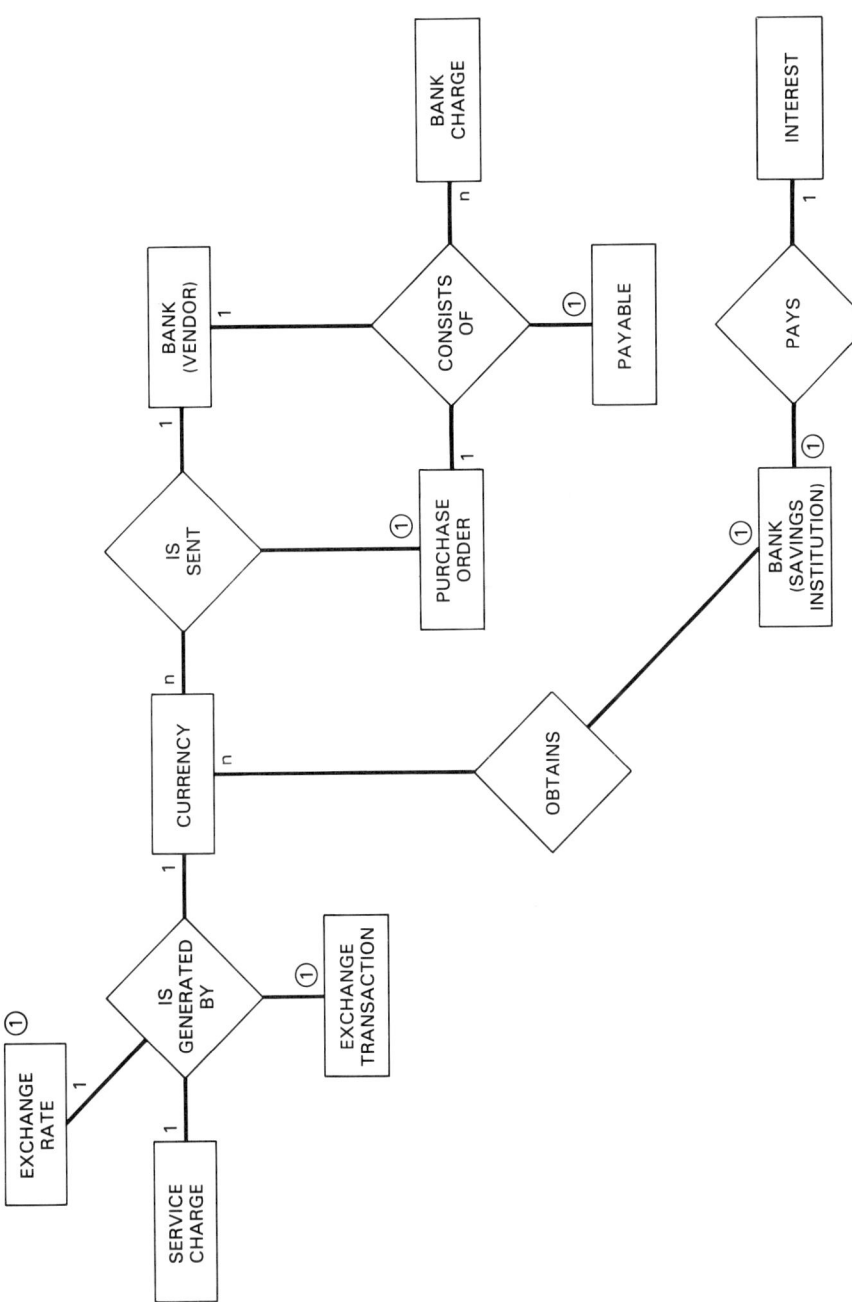

Figure 24: MaxiChange Currency Exchange System—Entity-Relationship Diagram, Part 1

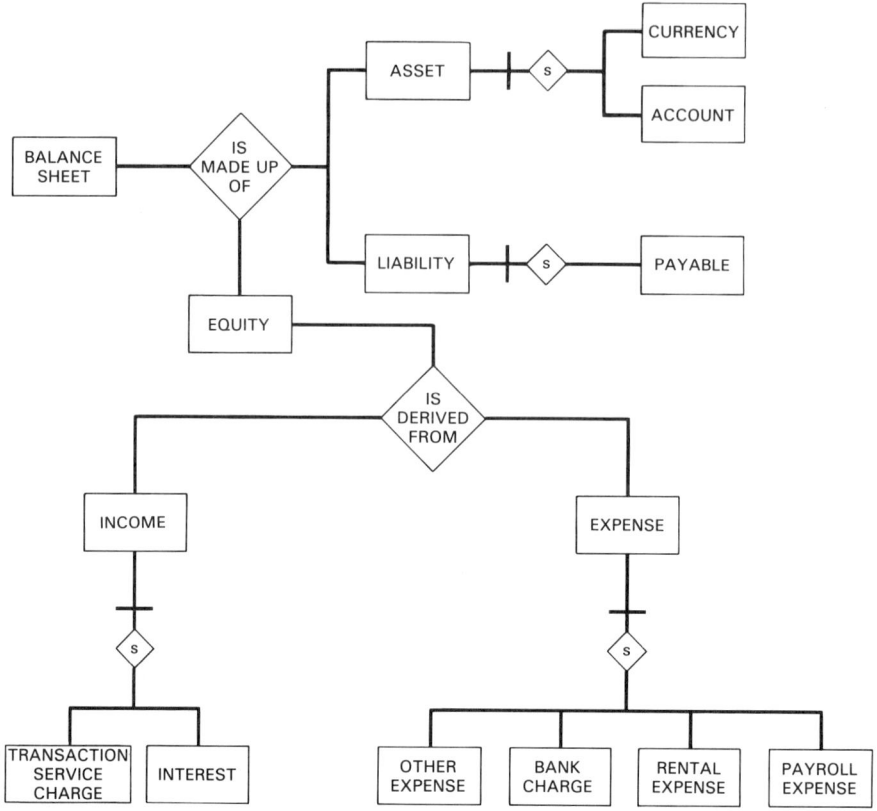

Figure 25: MaxiChange Currency Exchange System—Entity-Relationship Diagram, Part 2

ENTITY-RELATIONSHIP DIAGRAM POLICY NARRATIVE

An *exchange transaction* is generated by a single, fixed service charge and an *exchange rate* for each *currency* bought or sold.

A *bank* (savings institution) obtains excess *currency* from the exchange booth and pays *interest* on that excess *currency*.

A (money) *purchase order* for one or more *currencies* is sent to a *vendor* (bank) when the exchange booth needs to replenish money inventory.

Exchange booth *payables* consist of one or more *bank charges* owed to a *vendor* (bank) for *currency*, for each *purchase order*.

The *payable* is satisfied when the exchange booth pays the *bank charges* to the *vendor* (bank), *rental expense*, or *other expenses*.

Sec. 4.2   The Detailed Diagrams   **77**

> The exchange booth's *balance sheet* is made up of *assets, liabilities,* and *equity* (shareholders equity is not a consideration and is factored in by head office).
>
> *Assets* consist of *currency* available in the exchange booth and held in the *bank* (savings institution).
>
> *Liabilities* consist of *payables* for *currency* due to the *vendor* (bank) and other miscellaneous *payables*.
>
> *Equity* consists of *income* and *expenses*.
>
>> *Income* is derived from *interest* earned by excess *currency* held in the *bank* (savings institution) and from the *transaction service charge* when customers buy *currencies*.
>>
>> *Expenses* include *payroll expense, rental expense, bank charges* for acquiring *currencies* and administering accounts, and miscellaneous *other expenses*.

Given this basic information, where do you start? The blank page syndrome sets in and you may be tempted to get your programmers to start writing some code. But you know that wouldn't be productive, so you start from the beginning and create an *event list*. You assess the events that occur *outside* the system that it must respond to and time-related events that occur *inside* the system that it must satisfy. After analyzing the basic requirements, you arrive at the following list of potential events.

MAXICHANGE CURRENCY EXCHANGE SYSTEM
POTENTIAL EVENTS

1. Customer wants to buy currency
2. Customer wants to sell currency
3. Time to replenish currency
4. Time to deposit excess currency
5. Time to receive interest
6. Time to pay vendor bank charges
7. Time to pay rent
8. Time to pay employees (or employees want pay)
9. Time to pay other expenses
10. Market exchange rates change
11. Time to recognize new currency
12. Time to de-recognize currency
13. Government wants employment reports
14. Government wants payroll remittances
15. Time to report profit-loss to head office

While you are analyzing the potential events in the system, in the back of your mind you are thinking about some of the rules that might apply. For example, you discover that most events are invariably time-related, since most of the world operates on a specific time scale rather than in a state of anarchy. Payroll is issued at a specific time; currency must be replenished at that time when inventory dips below the reorder level; we get rid of excess money at that time when there is too much cash on hand; we pay rent at a specific time during the month; expenses are paid according to some time-related policy, such as 30 days after receipt of invoice; head office wants financials reported once a month; and so on.

The best way to arrive at a reasonably complete event list quickly (say, in a day or two for most nontrivial systems) is to create the list *with the user*. This will give the user a sense of ownership and participation, and the *innermost substance* of the system will be verified immediately as the work is being done.[8]

Working against your *event list* and using the Essential Information Model, we can then start drawing a mini-model for each of the events listed for the system. We do this without any initial concern for integration of functions or redundancy; nor do we consider the large amount of detail that is usually necessary to build in controls. The objective is to build a business specification quickly, accurately, and unambiguously, without being encumbered with the inordinate detail and complexity of the particular implementation technology.

Upon analyzing the first two events on the list—*customer wants to buy currency* and *customer wants to sell currency*—we discover that "buy" and "sell" are the opposite sides of the same thing; that is, the currency exchange booth is always seen to be *selling* currency, whereas the customer is always seen to be *buying* currency. With this new perspective, we combine the two events into one new event—*customer wishes to exchange currency* with a resulting mini-model that could look something like Fig. 26.

When we create this mini-model, however, we immediately beg a couple of questions: (1) What happens if we don't have enough or any of the currency requested by the customer, and (2) How do we keep track of demand (which is based on customer requests) so we will know when to recognize new currencies? So we very quickly amend the diagram to reflect this reality, with the result shown in Fig. 27.

Since we don't want to take forever and model the universe, we stop there and quickly go on to the next event on our list, *time to replenish currency*.

---

[8] IBM's joint application design method (JAD) works in a very similar manner, and reasonably well. However, the results tend to get a bit implementation-dependent when this approach is taken with a multitude of users rather than only the most senior people.

Sec. 4.2   The Detailed Diagrams   79

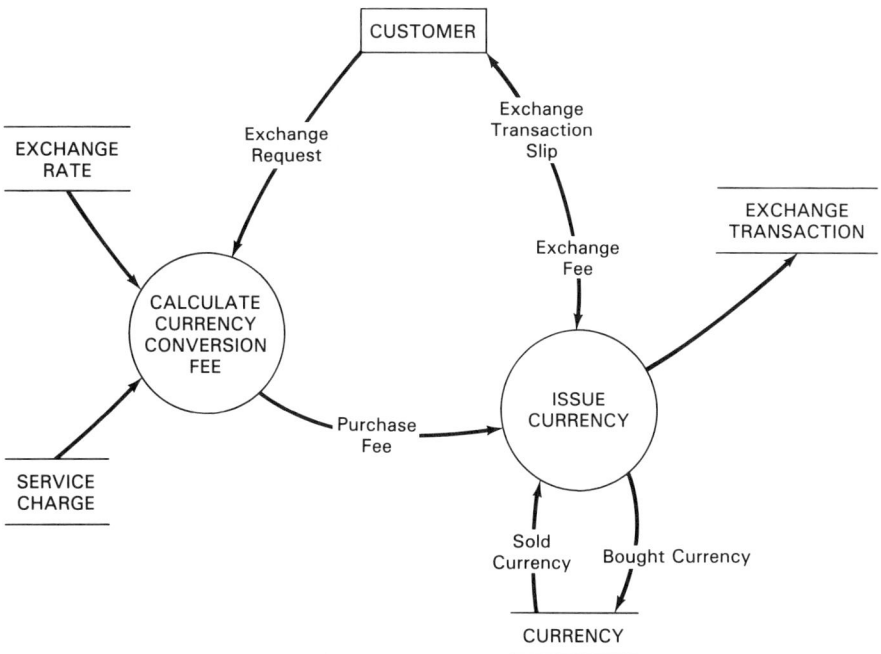

Figure 26: Customer Wishes to Exchange Currency—Event Model

Using the basic information from our Essential Information Model, we arrive at the mini-model diagrammed in Fig. 28.

To determine whether we have depleted our currency sufficiently to justify reordering, we query the CURRENCY logical data store and, if needed, we issue a currency purchase order to our vendor after logging the purchase order. Established reorder points must be balanced against CURRENCY DEMAND in case there have been recent changes in demand levels. At some point we receive a new supply of currency from our vendor with an invoice. To ensure that we actually ordered the currency, we check against our PURCHASE ORDER logical data store (PURCHASE ORDER VALIDITY QUERY). If we find we didn't order the currency, we "reject" to an undefined process. We can model the error handling later (remember, the idea is to model the *innermost substance* of the business, fast). If we can verify that we actually did order the currency, we increase our inventory in the CURRENCY logical data store, and we record the currency invoice as a PAYABLE.

One small point: the CURRENCY data store is purely *logical*; that is, it may not be implemented as a separate file in the final system implementation. It may form part of a database as a key attribute, or it may be a record. Also, CURRENCY clearly consists of *two* physical components if we think in terms of implementation. First, it is a mechanism to *record* data about currency (such as a logical database record); second, it is a *physical repository* for the actual cash (such as a cash drawer behind the cage). But, to keep our essential model uncluttered and understandable, we have chosen to represent both of these physical views as one logical data object. It helps to keep life uncomplicated and it wouldn't be helpful to illustrate both views in an Essential Process Model.

Our next event, *time to deposit excess currency*, is fairly straightforward. Again, we query our logical CURRENCY data store to see how much currency we have on hand. If the currency level is greater than the established amount for individual currencies, we record the fact that we are getting rid of some of it (CURRENCY REMOVAL NOTICE), keep track of the DEPOSIT, and send it off to the bank. (See Fig. 29.)

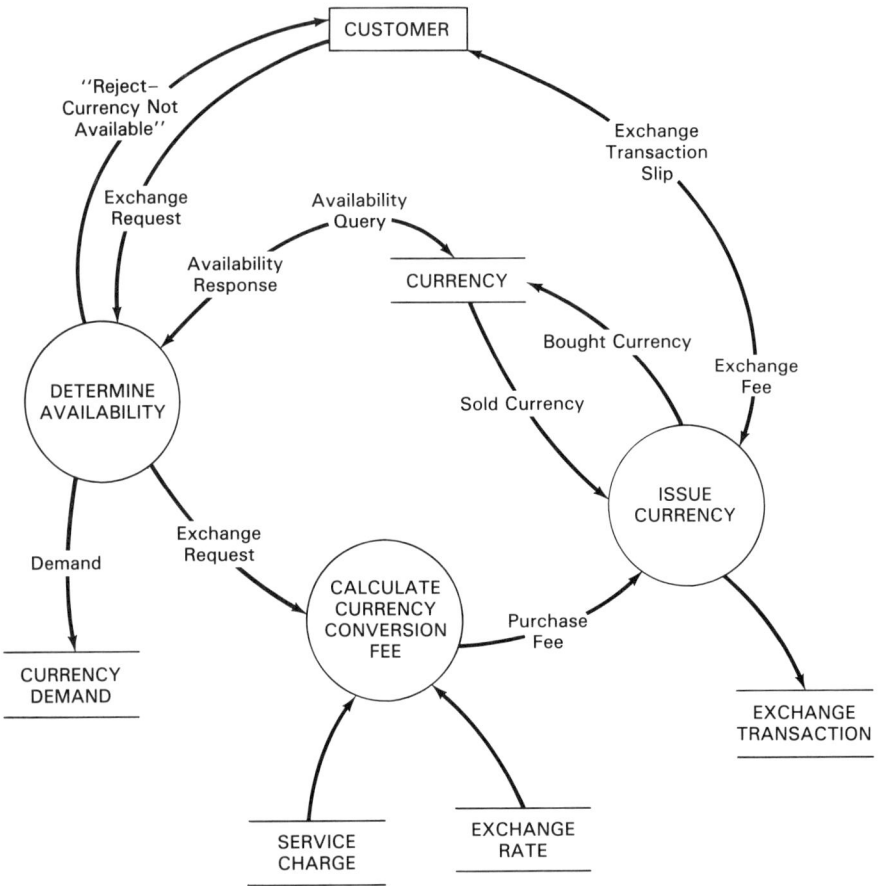

Figure 27: Customer Wishes to Exchange Currency, Revised—Event Model

Sec. 4.2   The Detailed Diagrams   81

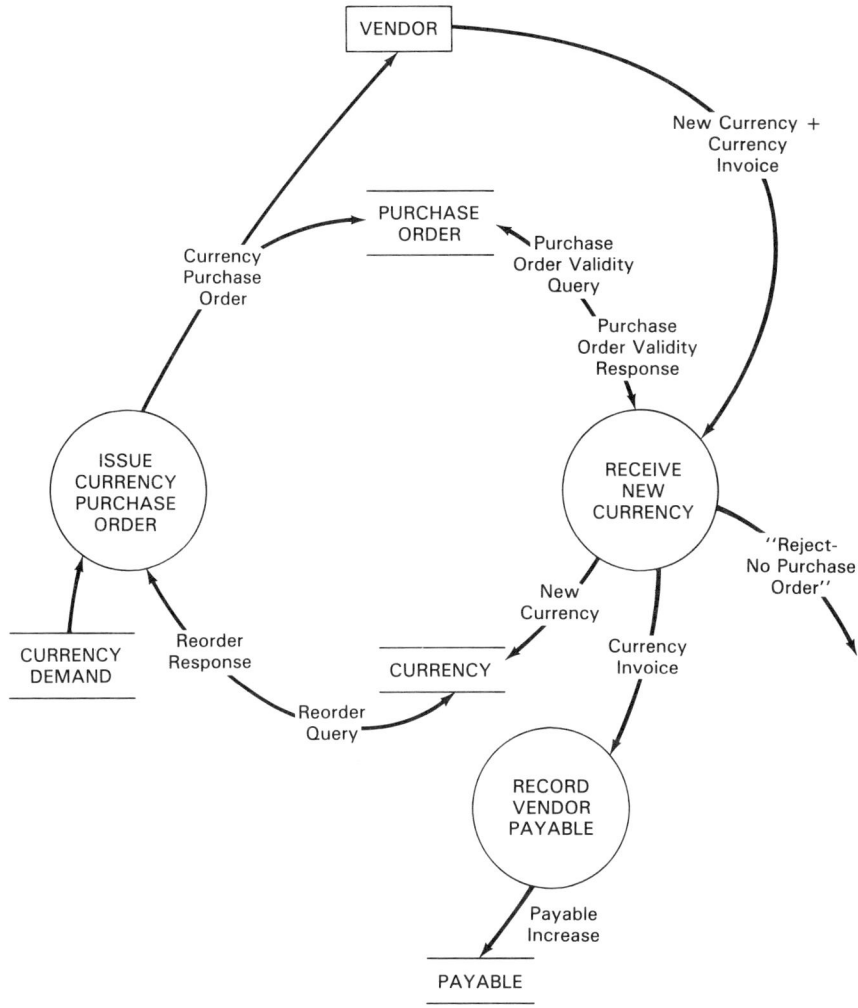

Figure 28: Time to Replenish Currency—Event Model

Our next event, *time to receive interest*, is in response to the preceding event, *time to deposit excess currency*.

From time to time the currency exchange booth receives a statement from the bank. (This may not be exactly the way it works in reality, but it will do for the purpose of our example.) We query our DEPOSIT record to make sure it matches what we actually did send to the bank, and we record INTEREST (see Fig. 30)

The next event on our list, *time to pay vendor bank charges*, give us our first real opportunity to directly relate the processes of one event-based model to the processes of another (see Fig. 31).

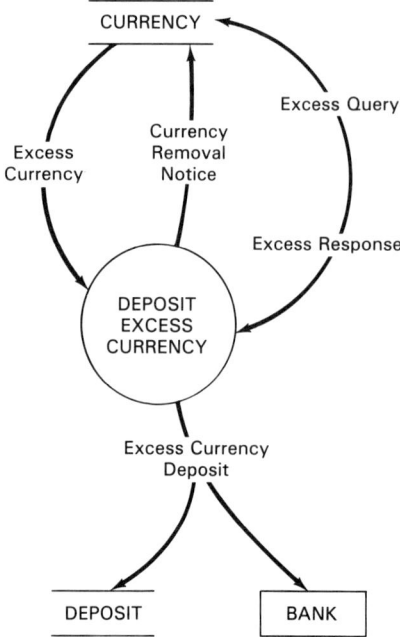

Figure 29: Time to Deposit Excess Currency—Event Model

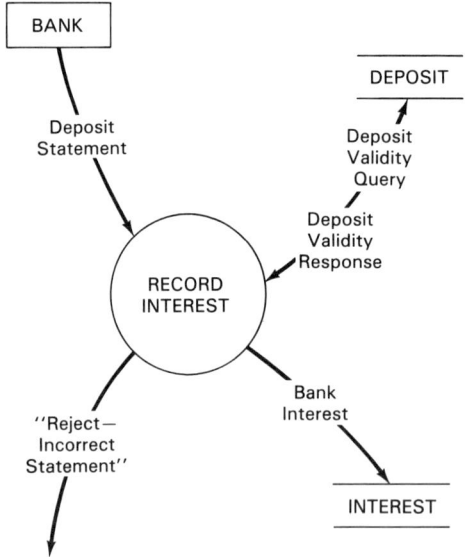

Figure 30: Time to Receive Interest—Event Model

Sec. 4.2 The Detailed Diagrams

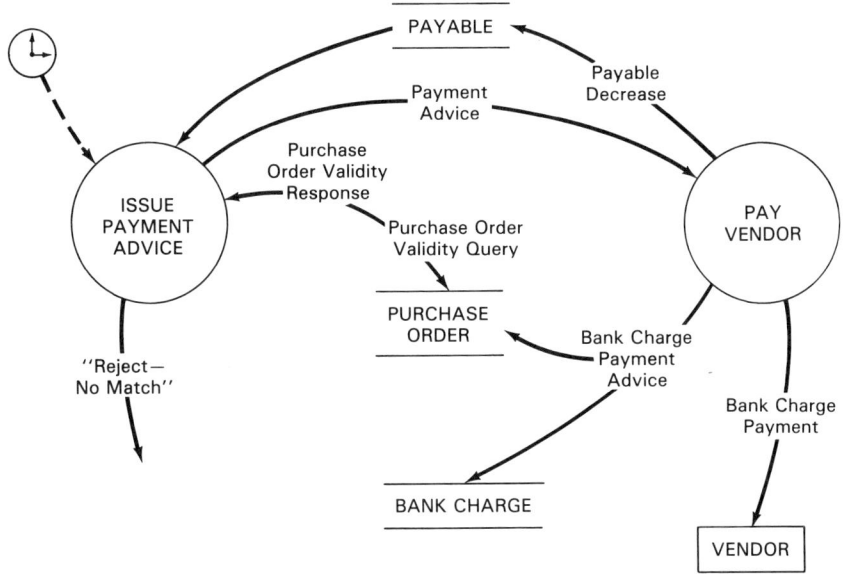

Figure 31: Time to Pay Vendor Bank Charges—Event Model

In Fig. 28 (*time to replenish currency*), you will remember that we recorded the vendor payable when we received a new supply of currency from the bank. Banks, having no sense of humor, naturally want us to pay for that purchase order at some point. Well, this is it.

After determining what payables are due, we reverify that the payable is for a legitimate purchase order (we did this also when we received the currency from the vendor). If it's okay, we issue a payment advice, decrease the payable, and amend the purchase order to reflect the fact that it has been paid. Additionally, we increase our BANK CHARGE record and send the payment to the vendor. The details of how we do this, and all the specific data we need to collect, can easily be worked out when expanding the *essential model* into the Functional System Model or when creating the Design Blueprint.

The little clock pointing into the process ISSUE PAYMENT ADVICE is only intended to indicate clearly that time is a stimulus to this process. We use it regularly. You might also find it a useful tool for communicating with users when walking through your data flow diagrams.

About once a month, it's *time to pay rent*. In our example (Fig. 32) we decided (after discussing it with Horacine Clutch) that rent should be recorded as any other payable, even though we don't receive an invoice each month. (We do receive adjusting invoices annually, however, for taxes, common area charges and maintenance, but we decided to leave that for expansion in the Functional System Model.) Therefore, we obtain the amount from the PAYABLE data store, decrease the amount for the time

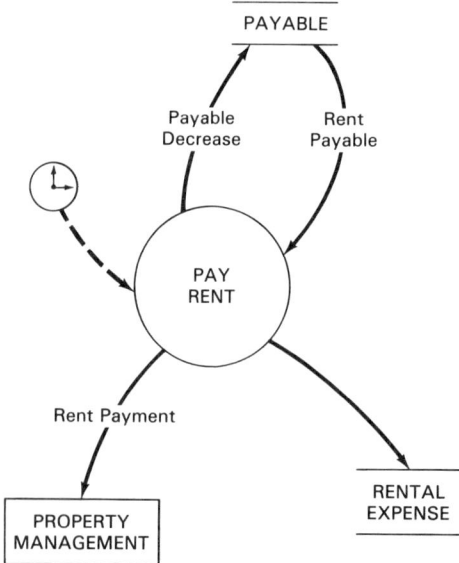

Figure 32: Time to Pay Rent—Event Model

period affected, record the RENTAL EXPENSE, and issue the payment to the Property Management group.

Time to pay employees, the next event, has intentionally been kept very simple so our example wouldn't get out of hand. Payroll is just not a simple affair anymore, but Fig. 33 at least reflects the *substance* of what we

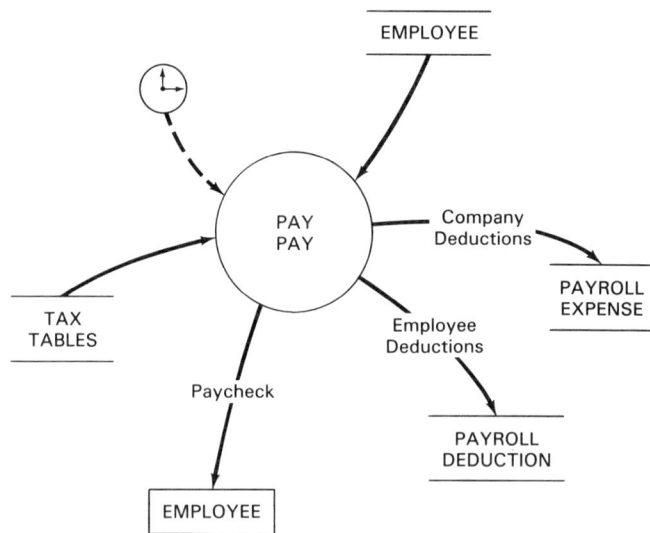

Figure 33: Time to Pay Employees—Event Model

Sec. 4.2 The Detailed Diagrams

need to do. We decided to keep PAYROLL DEDUCTION a separate key object, being *employee* deductions, from PAYROLL EXPENSE, which is the *employer's* (company) contributions.

The mini-model for *time to pay other expenses* (Fig. 34) is almost the same as the model for the *time to pay vendor bank charges* event. The only things that change are a couple of data flow names, the identification of the terminator, and the name of a logical data store. This kind of similarity between some event models should be kept in mind for when we *normalize* these models into a cohesive, partitioned Essential Process Model.

The next event, *market exchange rates change*, could look like Fig. 35.

To bring *Currency Type* in from the CURRENCY logical data store may not be necessary, depending on how this segment of the system will be designated physically. For example, *Market Exchange Rates* received from the money market could conceivably be *all* rates, rather than just the few the exchange booth is working with today. For the extra few nanoseconds, why not? It may make design and construction a lot easier, and it may make the system more flexible and adaptable in the future. In the meantime, it does no harm to include *Currency Type* so that the opportunity of using it this way is at least available to the users and designers. You are allowed to change your mind, you know.

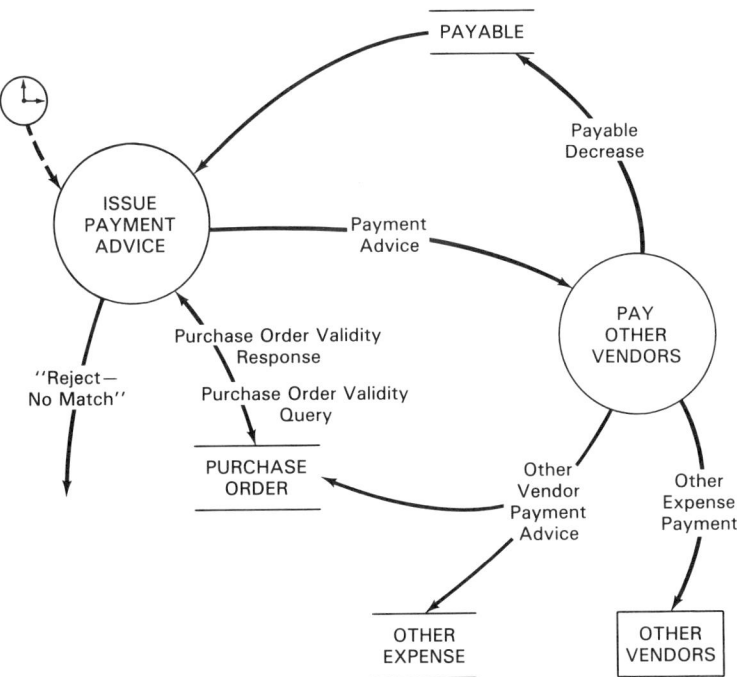

Figure 34: Time to Pay Other Expenses—Event Model

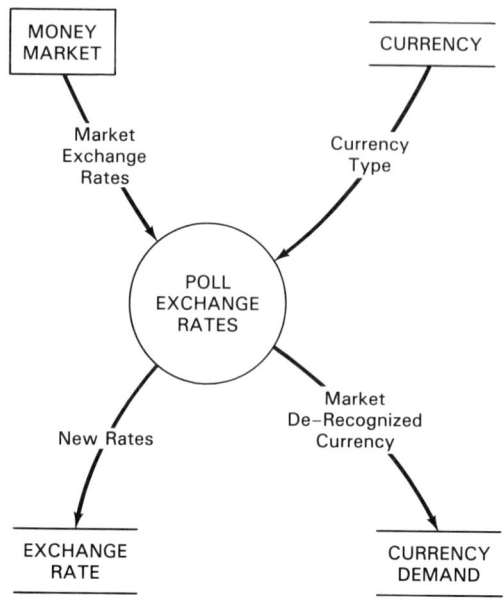

Figure 35: Market Exchange Rates Change—Event Model

The other interesting side to this mind-model is that *Market De-recognized Currency* is captured on the logical store CURRENCY DEMAND. You will remember that we used CURRENCY DEMAND in Fig. 27, *customer wishes to exchange currency*. In that process we were collecting demand statistics when the exchange booth received an exchange request, statistics for both currencies the booth had in inventory and currencies not handled by the booth. Two other events are also related to currency demand, (1) Time to Recognize New Currency and (2) its flip side pal, Time to De-Recognize Currency, neither of which we have modeled yet. Both of these events will need data collected on CURRENCY DEMAND. The second event, Time to De-Recognize currency, can be driven by either of two scenarios: (1) a currency must be de-recognized because there is no meaningful demand by customers, and (2) the money market no longer recognizes the currency; therefore, by default, there is no demand. Since money market demand is based on bid and ask prices, and no demand means no bid and ask quotations, we decided to place such "de-recognized" currency on the CURRENCY DEMAND logical data store. Currencies without bid and ask prices could also be placed on the EXCHANGE RATE record, but we felt that de-recognized currencies were conceptually different from currencies with bid and ask prices of greater than zero. How this is *physically* implemented, however, is a design consideration and can be addressed later in the development process.

Which brings us to the *time to recognize new currency* event (Fig. 36).

Sec. 4.2 The Detailed Diagrams   87

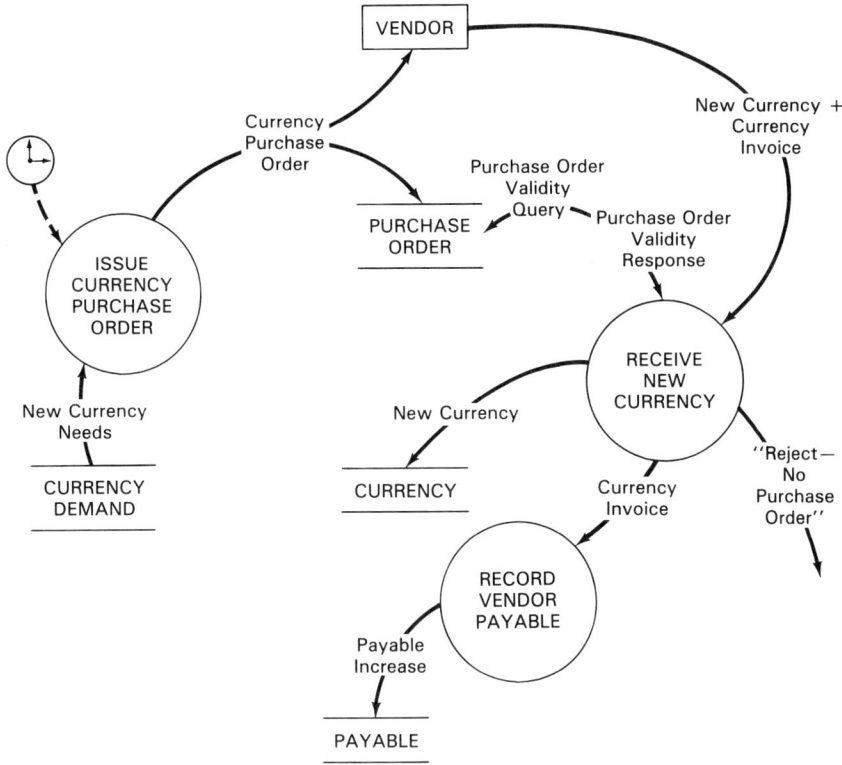

Figure 36: Time to Recognize New Currency—Event Model

Based on CURRENCY DEMAND, a purchase order for new currency is sent to the vendor bank and recorded. The rest of this process is surprisingly similar to Fig. 28, *time to replenish currency*. You may want to go back and read that one again. This similarity should again be kept in mind for when we get to *normalizing* the model and developing a cohesive, integrated view of the system's functions.

The flip side of recognizing new currency is *time to de-recognize currency* (Fig. 37).

Remember that we collected customer demand statistics for currencies in the *customer wishes to exchange currency* event? Well, now it's time to do something with this data.

When our customers are no longer interested in certain currency, we naturally want to stop dealing in it. Based on some predefined algorithm, the system determines which currencies are no longer in demand. It also looks for currencies without bid and ask prices, as provided by the money market (see our discussion on Fig. 35, *market exchange rates change*). Currencies that are no longer in demand must be removed from inventory, if there is any left. So a *Currency Removal Notice* is sent (electronic or paper, whichever

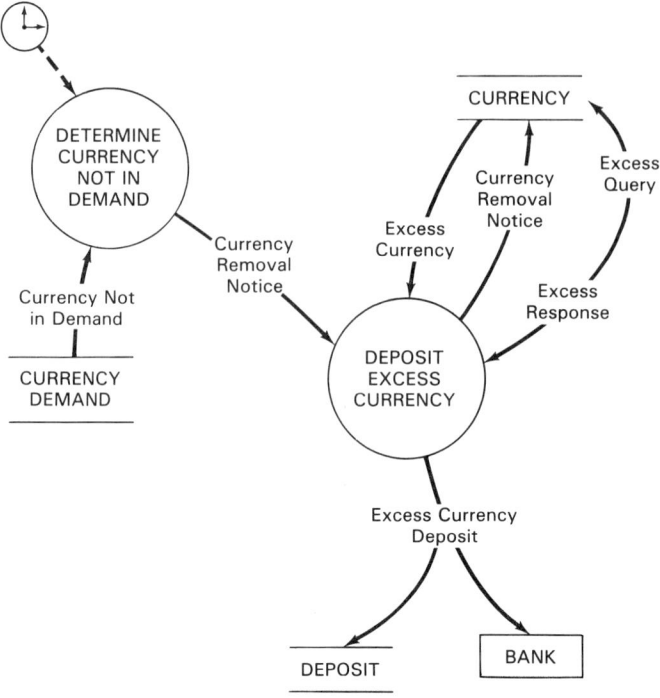

Figure 37: Time to De-recognize Currency—Event Model

is appropriate in the physical implementation) to the Deposit Excess Currency function. This process checks against the logical CURRENCY record to see whether there is any currency (to be removed) in inventory (*Excess Query* and *Excess Response*). If there is something in inventory, the currency to be removed is annotated with a *Currency Removal Notice* and the *Excess Currency* is removed from inventory, recorded (DEPOSIT), and shipped to the bank. Interestingly, the Deposit Excess Currency function is exactly the same as its counterpart in Fig. 29, *time to deposit excess currency*.

There may be other ways of providing a *Currency Removal Notice* to the system. For example, we might look at the process in Fig. 38 instead. By taking this logical view, we *maximize cohesion* and *minimize the coupling* of processes, and it is therefore probably the better view. The only real change is that the processes communicate through the logical record *CURRENCY* rather than directly, and an audit stamp takes the place of the currency removal notice illustrated in Fig. 37.

Figures 35 to 38 now give us cause to go back and modify our Essential Information Model, or entity-relationship diagram. Remember how it looked, as in Fig. 39.

The information carried by the original entity-relationship diagram reflects the information important to head office. It does not include

Sec. 4.2   The Detailed Diagrams   89

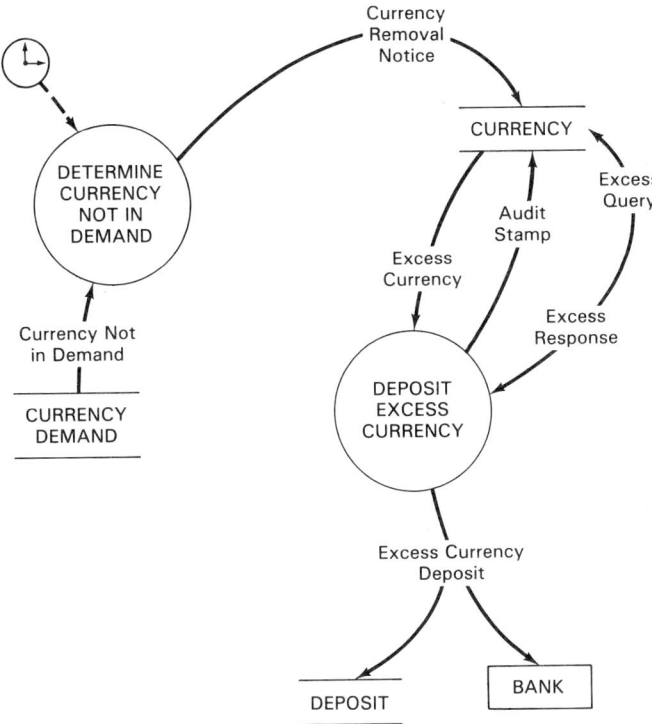

Figure 38: Time to De-recognize Currency—Event Model—A Second View

information pertinent to the actual logical operation of the system itself—that is, additional data needs *not of interest* to head office and the relationships among the data objects. But as we developed mini-models for the events we *discovered* other needs and relationships, all of which will influence our final database design. The (partially) amended entity-relationship diagram might now look something like Fig. 40.

Always bear in mind that an entity-relationship diagram is supposed to define objects or *things of interest* to the business. The questions, then, always become, Who's view of the business, and what is it that interests them?

This entity-relationship diagram will also be amended around the area of the PAYABLE and PURCHASE ORDER objects as a result of discoveries when creating the Essential Process Model. Also, when creating that high-level entity-relationship diagram (Essential Information Model), we will probably discover changes we need to make to our Essential Process Model as a result of the data attributed to the various objects. Do you remember the diagram shown in Fig. 41?

When we discussed this, we said that the development of the Essential Information Model and the Essential Process Model provides knowledge to

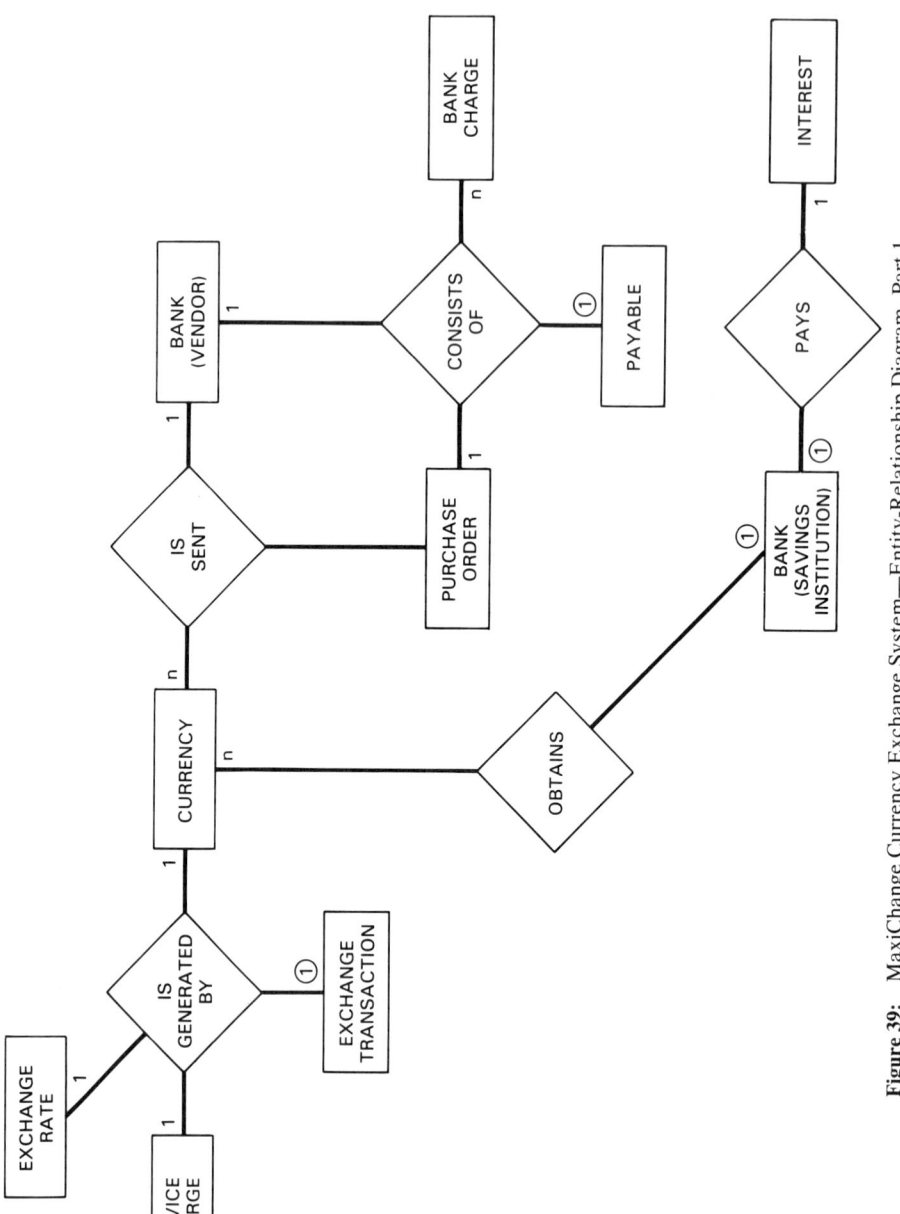

Figure 39: MaxiChange Currency Exchange System—Entity-Relationship Diagram, Part 1

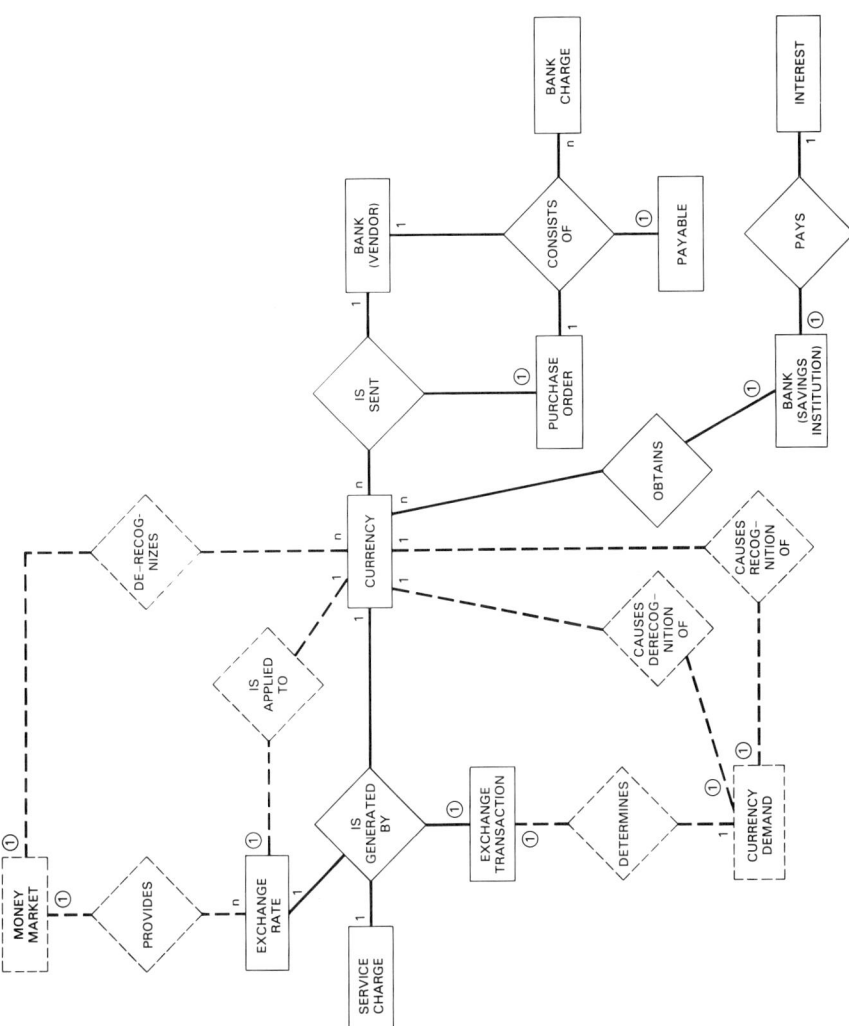

Figure 40: Partially Amended Entity-Relationship Diagram—MaxiChange Currency Exchange System

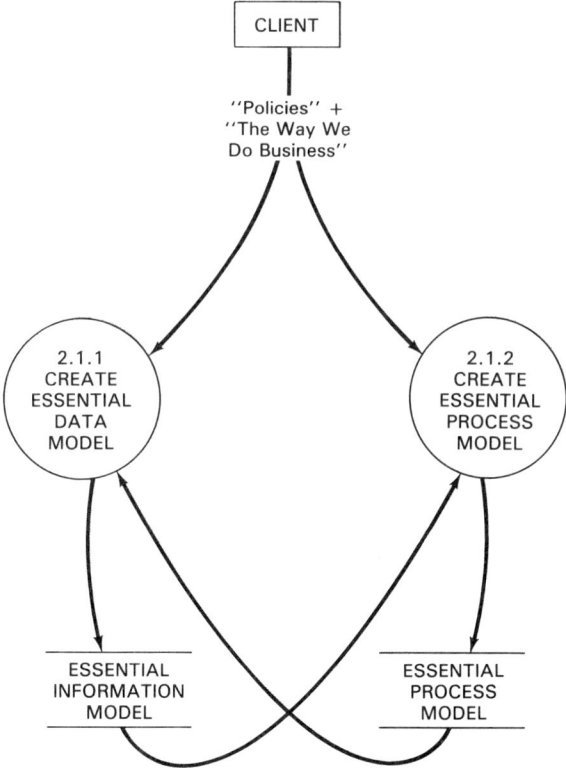

Figure 41: Create Models of Essence (2.1)—Partial

each model so that the result could be an unambiguous, concise, predictive model of the user's business needs and functionality. You will remember that we said the information model was a *knowledge machine*, while the process model was an *output-generating machine*. The information model tells the process model what knowledge it has *available* to produce output, whereas the process model advises the information model what knowledge it *needs* to produce the desired output. One without the other will invariably lead to a lot of guesswork ("Gosh, I wonder if we have the necessary data?"), and a less than wonderful database design ("Well, we can put this stuff in here, as part of the Master File!").

Having tackled the relationship between the Essential Information Model and the Essential Process Model, let's return to completing the mini-models for the events in our system and try to get Ted a couple of weeks of vacation in the summer.

We have intentionally left the modeling of the remaining events at a considerably high level of abstraction, since any additional detail might cause you to fall asleep.

Sec. 4.2   The Detailed Diagrams   93

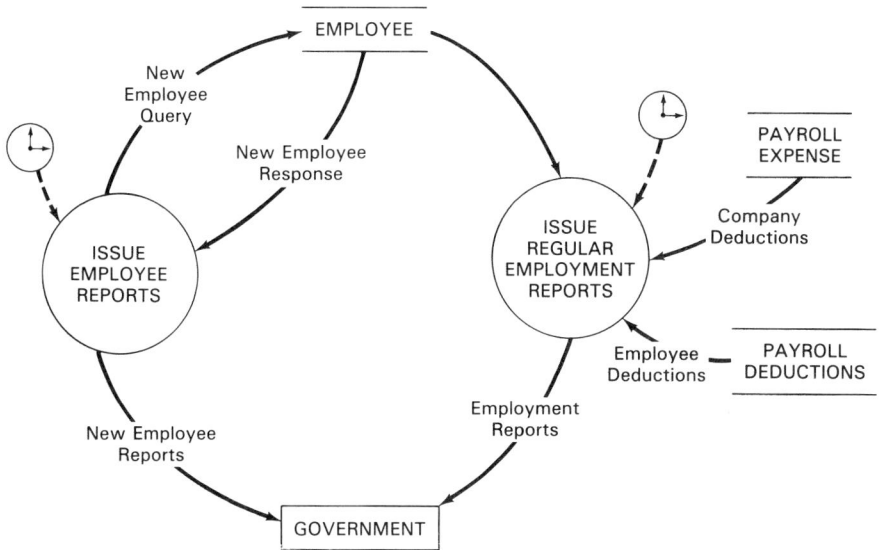

Figure 42: Government Wants Employment Reports—Event Model

Government wants employment reports (Fig. 42) could also be time-driven, for example, *time to issue government employment reports*. With either view, there is no change to its substance. New Employee Reports could also include such things as EEOC reporting, but we haven't included such detail here.

The next event, *Government wants payroll remittances*, (Fig. 43) (which again could be a temporal event) is an administrative burden of considerable detail carried by all companies. But we will pretend it is trivial in this case.

Finally we come to *time to report profit-loss to head office*. In his memo to Ted, Verbal Funderbuck, vice president of new business, said: "Galactic Headquarters wants profit-loss and balance sheet data on a monthly basis; this is to be uploaded from the local PC to the DataSchwartz 2938 mainframe at head office; head office will design its own reporting mechanism and will be responsible for reconciliation." Since we don't have much to do except collect the data the system has been gathering all along and send the data off to head office, our event-based model might look like that in Fig. 44.

Since a mini-model for each event identified in the system has now been completed, and all have been walked through with the user and approved, the next step is to *normalize* these independent processes. That is, we want to merge these individual processes into a larger, cohesive model, eliminating redundancy and merging common data flows and

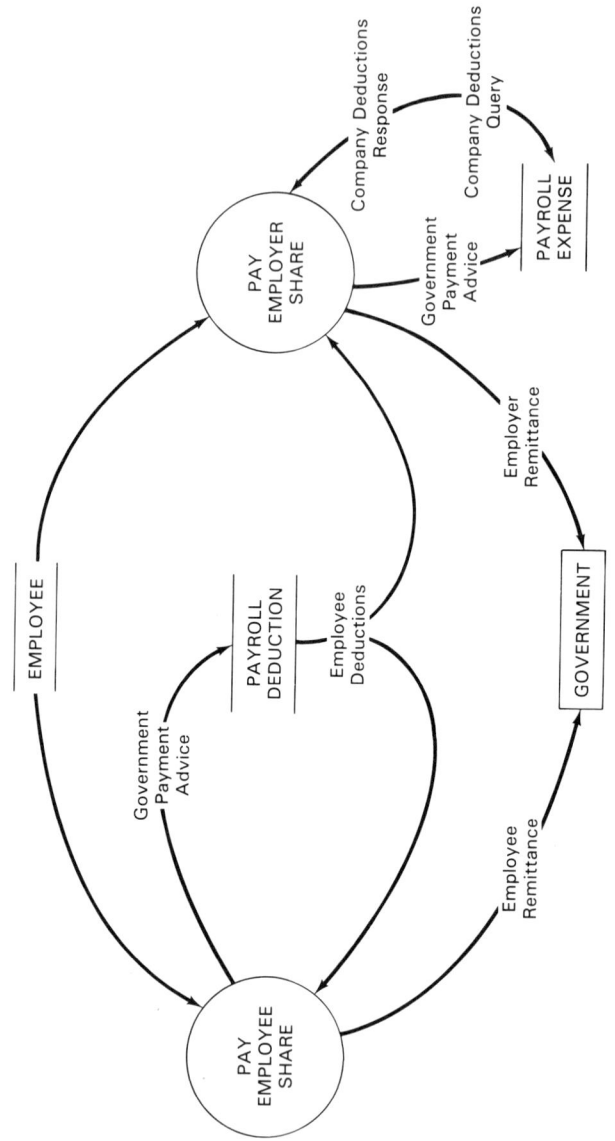

Figure 43: Government Wants Payroll Remittances—Event Model

Sec. 4.2   The Detailed Diagrams   **95**

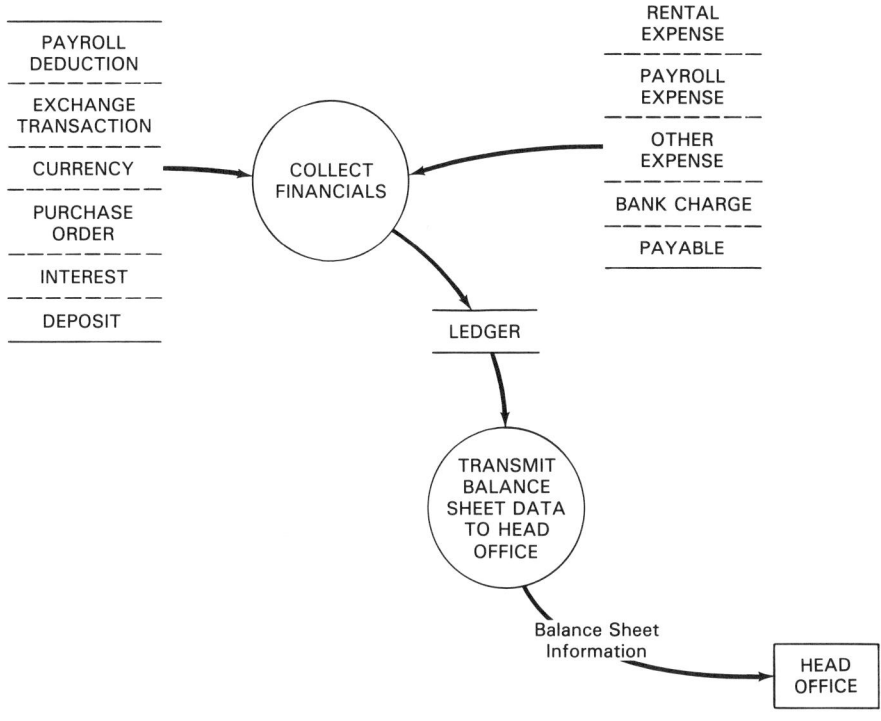

Figure 44:   Time to Report Profit-Loss to Head Office—Event Model

processes.[9] To be completely accurate, of course, we need a concise data dictionary at this time, as well as a descriptive narrative for each process. But for the sake of this example, let's bridge that gap intuitively (which is also sometimes fair!).

After analyzing our event models, we come to the conclusion that there are only five basic functions to this system: (1) *Exchange Currency*, (2) *Maintain Currency Inventory*, (3) *Pay Expenses*, (4) *Issue Payroll*, and (5) *Issue Head Office Information*. Arguably, *Issue Payroll* could be part of *Pay Expenses*.

> [9] A system is not merely a collection of processes like *Determine Availability*, *Poll Exchange Rates*, *Deposit Excess Currency*, and so on. A system would not work at all unless its processes were linked to one another by a network of connections through a database of information. In *The Society of Mind* (New York: Simon and Schuster, 1986), Marvin Minsky wrote: ". . . you couldn't predict what would happen in a human community from knowing only what each separate individual can do; you must also know how they are organized—that is, who talks to whom. And it's the same for understanding any large and complex thing. First, we must know how each separate part works. Second, we must know how each part interacts with those to which it is connected. And third, we have to understand how all these local interactions combine to accomplish what the system does—as seen from the outside." While Minsky wrote about the brain, our view of non-human systems is virtually the same.

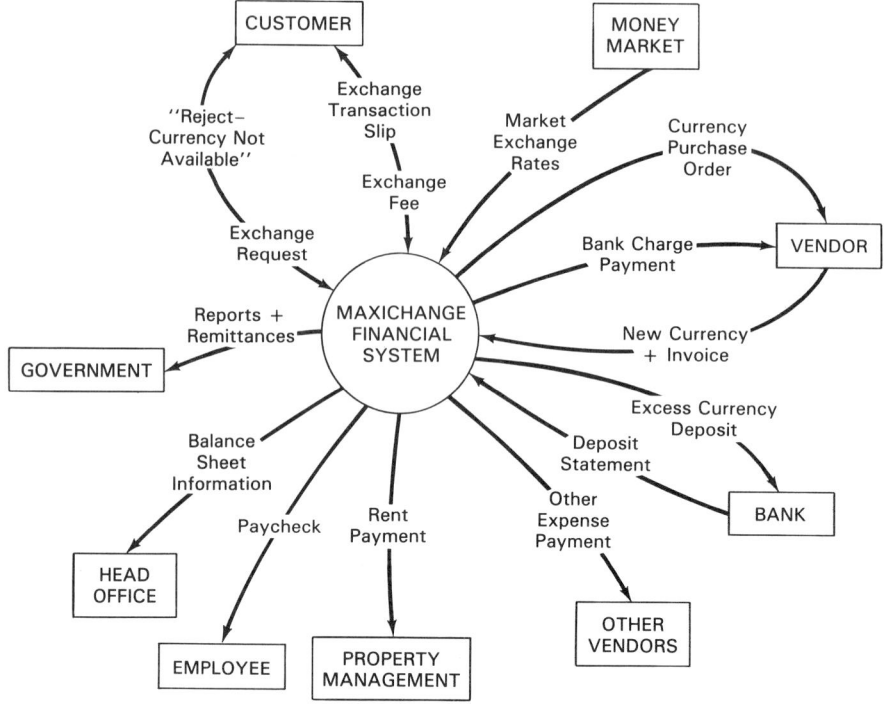

Figure 45: Context Diagram—MaxiChange Financial System

After looking for all the terminators on our event models, the context diagram will look like Fig. 45.

To develop the event list and event-based models, we worked *top-down*. To normalize these models we work from our event models *bottom-up*. When we later expand this Essential Process Model, we will again work from *top-down* within each of the processes identified. After combining the event models, eliminating redundancy, and merging common data flows and logical data stores, we have the result shown in Fig. 46.

Each of the bubbles (i.e., processes) in Fig. 46 has a corresponding lower level diagram—Figs. 47–51—to illustrate additional detail. Study them carefully to see how the preceding event-based models were normalized into a cohesive Essential Process Model.

Because this book is generally not about analysis, we normally would not go into this level of detail on *how to do* certain kinds of analysis; but we felt it was important enough to devote some time and space to it, so that you could get on a fast track quickly.

Also, there is no perfect model; the idea is to get an unambiguous understanding of the user's functional requirements so that we can get on with implementing the system as quickly as possible. This example is clearly

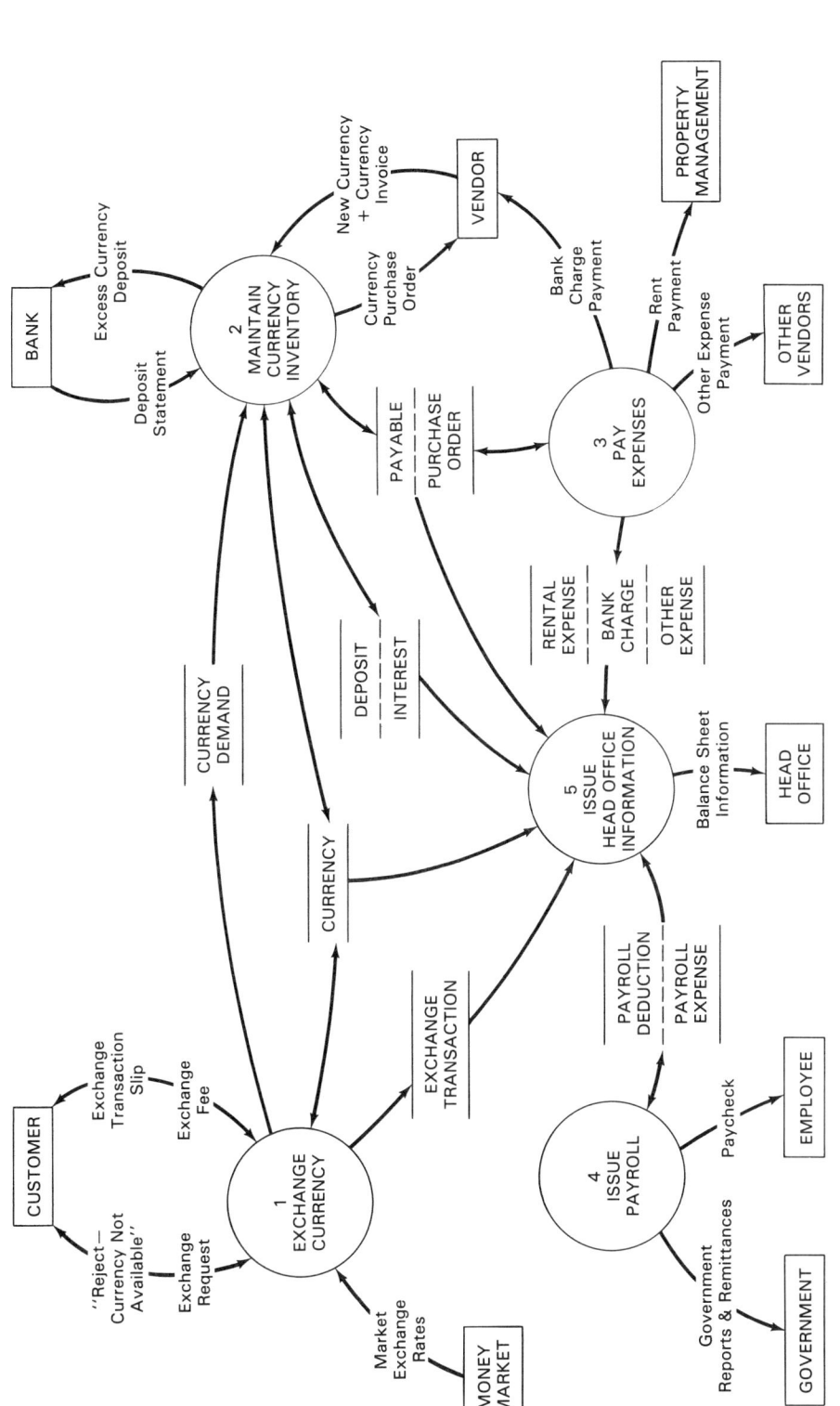

Figure 46: Diagram 0—MaxiChange Financial System

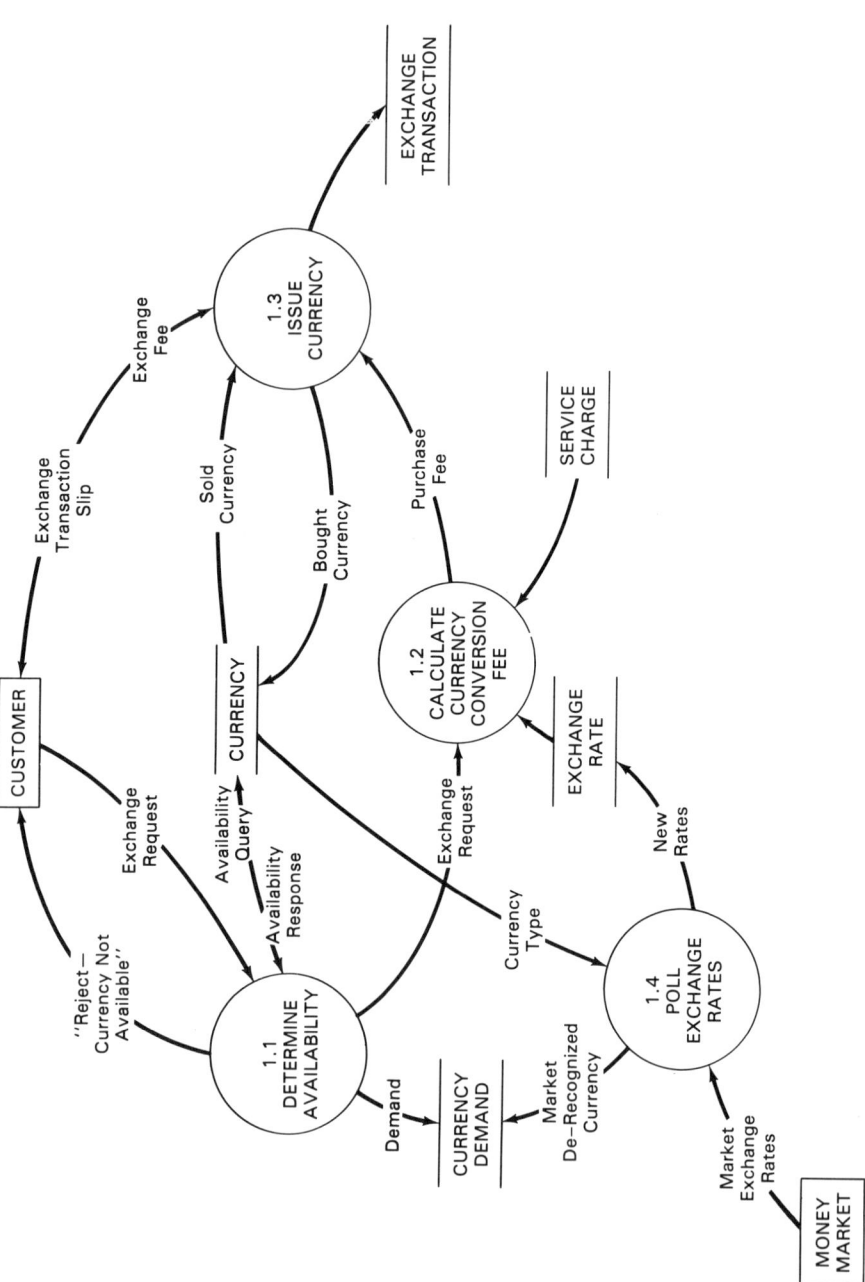

Figure 47: Exchange Currency (1)

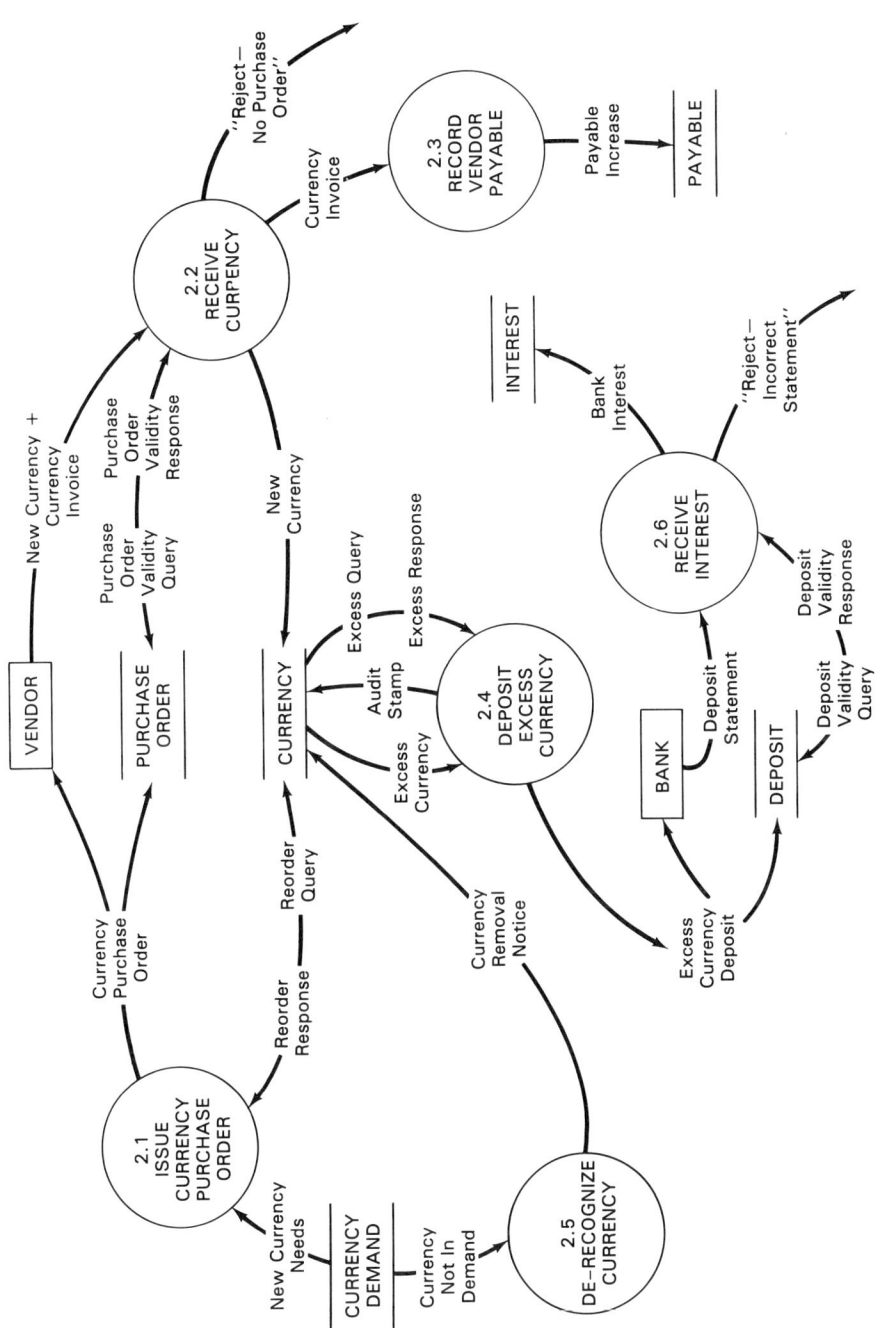

Figure 48: Maintain Currency Inventory (2)

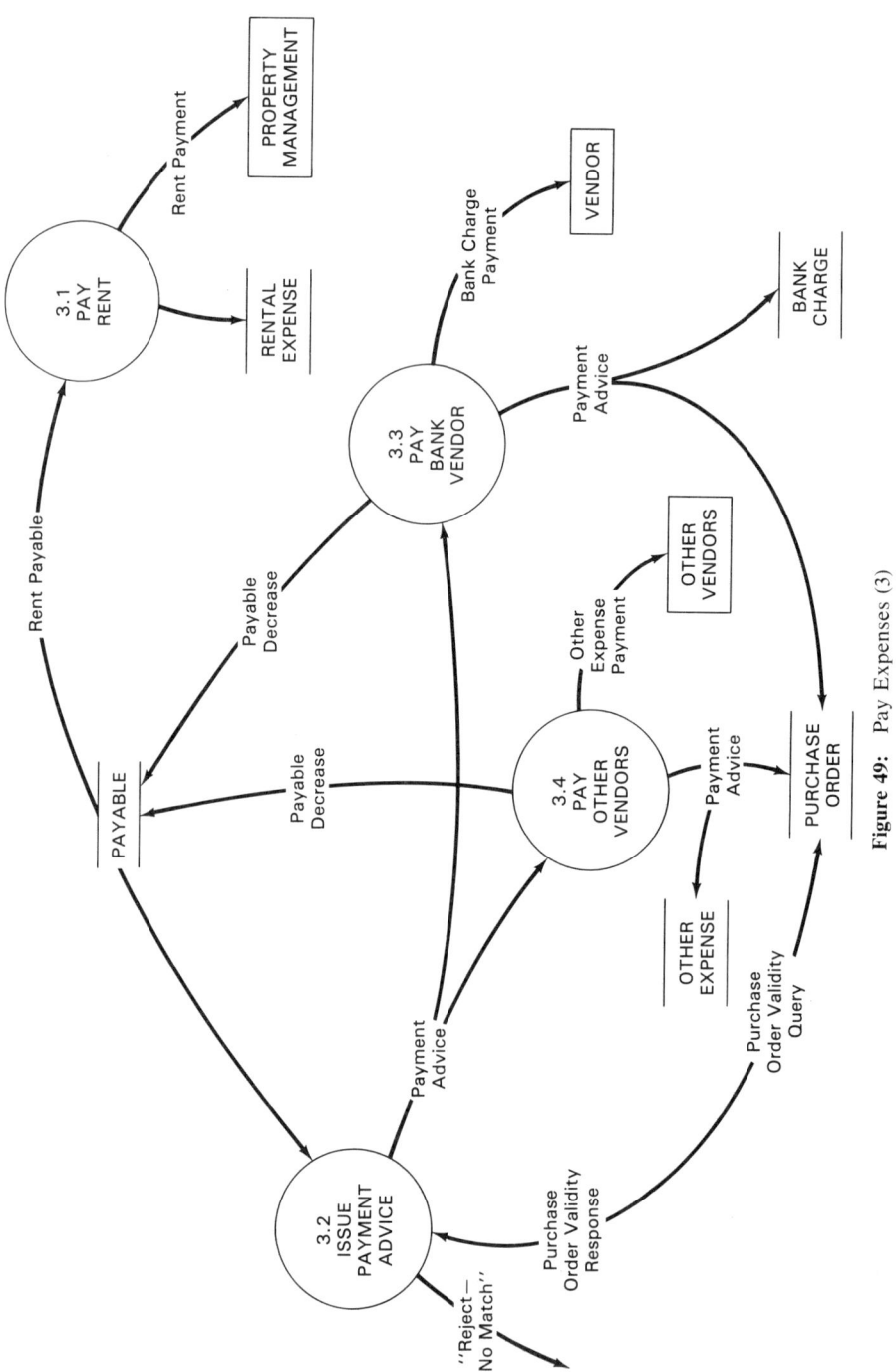

Figure 49: Pay Expenses (3)

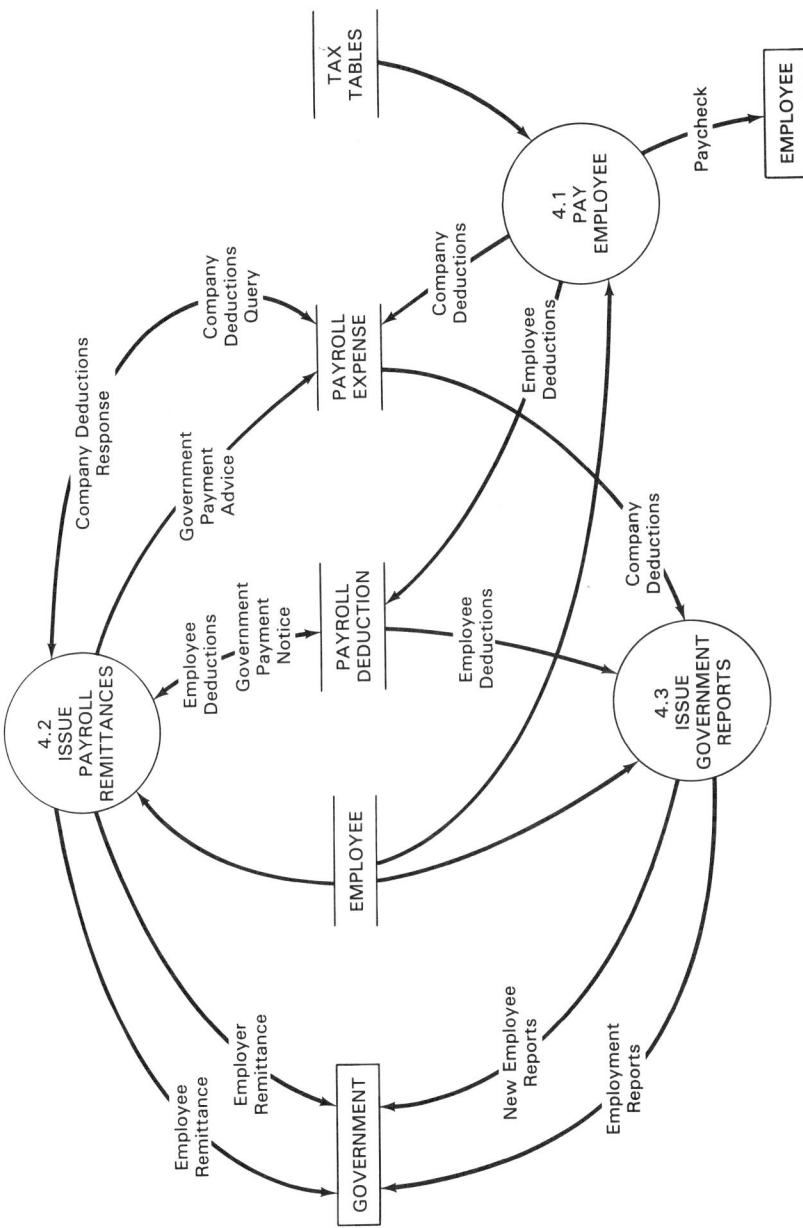

Figure 50: Issue Payroll (4)

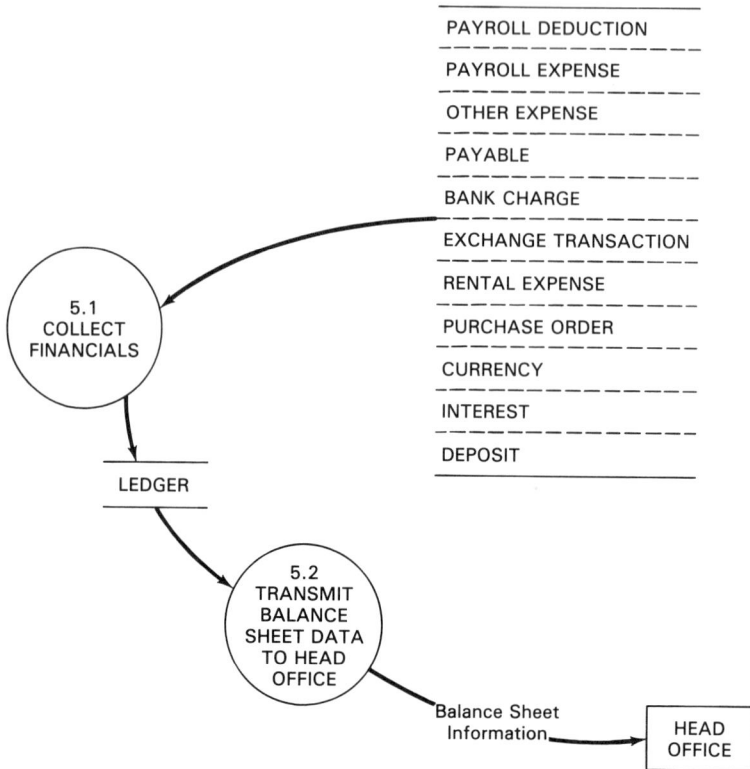

Figure 51: Issue Head Office Information (5)

not as detailed as it could be, nor is it as correct as it could be (there are a lot of correct views, but some are decidedly wrong too). But it is intuitively obvious what is going on, it is what the user wants, it was fast (less than a day), and it afforded a logical, cohesive view of the system (without burdening it with technology or controls), thereby permitting a better, clearer target for design. It's fast; it works; try it. With this fast start, perhaps Ted "Rambler" Yento will not be relegated to the company's life insurance subsidiary after all.

Refining Objectives and Constraints

Process 2.2, *Refine Objectives & Constraints* (Fig. 52), is an example of a situation that will occur throughout the development process.

This deliverable defines the preliminary scope of the project, as seen by the client. It includes assumptions, expectations, and priorities. It also includes constraints such as company and government policy, hardware and software, staff constraints, scheduling and target dates, budgets, and data

Sec. 4.2   The Detailed Diagrams   **103**

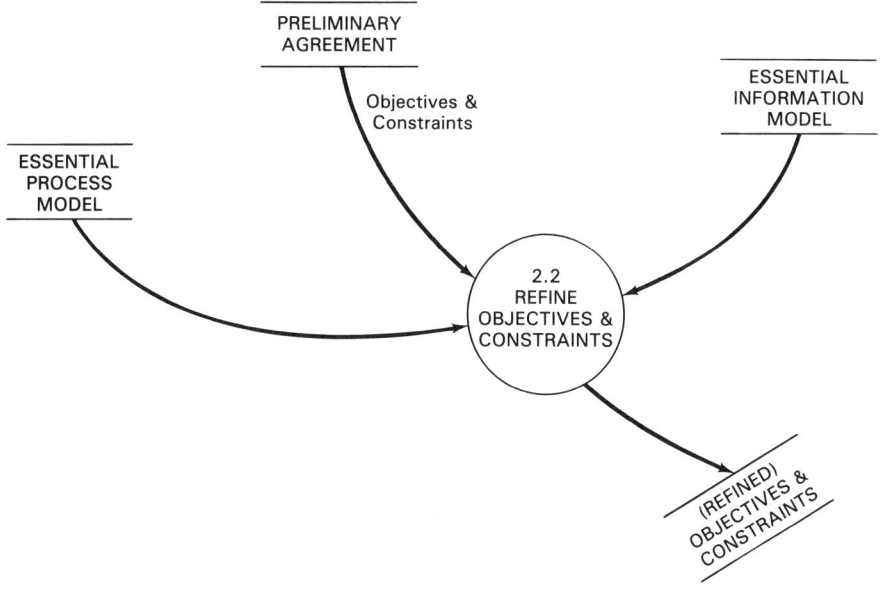

Figure 52:   Refine Objectives & Constraints (2.2)

privacy rules. It should also define system criticality (i.e., risk) and what will *not* be done by the development team.

Not only is Objectives & Constraints often modified as the project progresses, since all of these questions can't possibly be answered with certainty this early, but many other deliverables also go through the process of continual refinement. Many deliverables from a previous sphere of activity will be reviewed and enhanced in light of what is learned during the development process.

If it's important to keep an earlier version of a deliverable around, then the data flow name must change to reflect the continued existence of the previous document. For example, the Essential Business Model is eventually expanded (not redeveloped) to become the Functional System Model, but we want to preserve both versions, so we use different names to reflect their content and nature. Other deliverables (such as Objectives & Constraints, Version Plan, Cost-Benefit Statement, Acceptance Criteria, and others) are retained *only in their most current version* through the development process. If any of the documents are text, they can be stored on a word processor and very simply revised and expanded as work progresses.

Preparing the Cost-Benefit Statement

The client is paying the bills for this project, and she must have the information available to her at all times that will permit her to decide

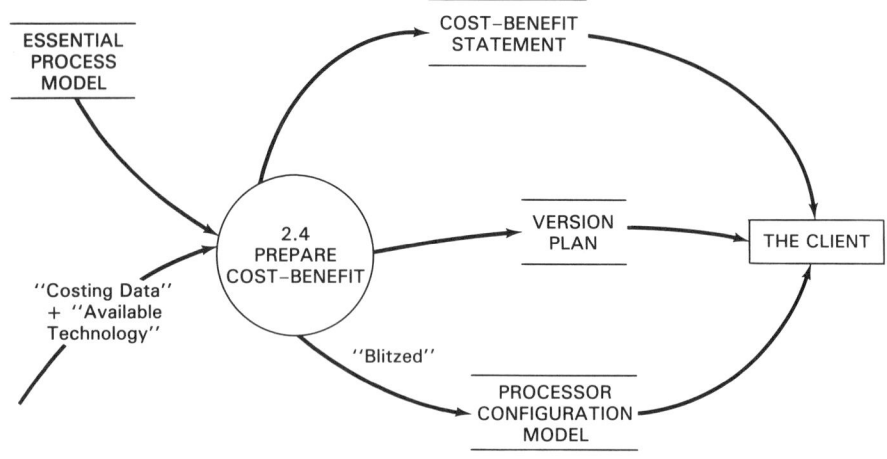

Figure 53: Prepare Cost-Benefit (2.4)—Level 1

whether to carry on or pull the plug. Certainly with the completion of the Essential models, some idea of the potential net cost of proceeding must be made available. Process 2.4 (*Prepare Cost-Benefit*), Fig. 53, doesn't, by itself, tell us very much about how we can go about getting some meaningful information to the client.

What it does show is that by using the Essential Process Model we've created, and some additional information we've called "costing data" and "available technology," we're able to put together a Version Plan, a Cost-Benefit Statement, and a tentative automation boundary by allocating the processes of the Essential Process Model to *processors* (machines and people), resulting in an early version of the Processor Configuration Model. It may seem odd to be talking about automation boundaries and processors during the creation of the Essential Business Model, but a look at what's going on at the lower level may help explain (see Fig. 54).

In order for the Development Team to make their first guess at possible costs, benefits, tentative processors, and an automation boundary, we must use the Essential Process Model to arrive at a draft Processor Configuration Model. We can't estimate the cost of implementing a process, even roughly, if we're not prepared to guess whether it's manual or automated and what kinds of processors may be allocated. In fact, a variety of processors and person/machine boundaries should be attempted in order to provide alternatives in the Cost-Benefit Statement.

Within each tentative person/machine and processor boundary, a Version Plan should be produced. This deliverable will show a partitioning of the functions to be delivered to the client, such that each segment can be implemented independently, gradually giving the client an increasing level of

Sec. 4.2 The Detailed Diagrams

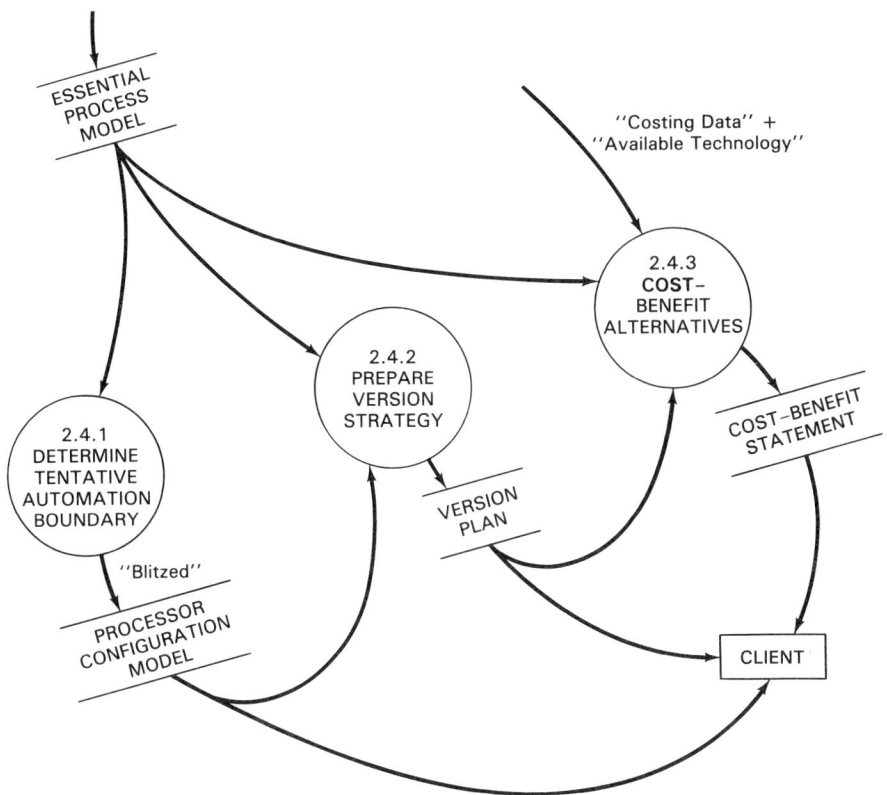

Figure 54: Prepare Cost-Benefit (2.4)—Level 2

support until the entire system is installed. Each segment (i.e., version) to be implemented will have a description of the functions included, an estimate of the schedule of delivery, and probable costs. A Version Plan conveniently permits you the luxury of delivering parts of the system separately, in a "phased" implementation, that is, with planned progression. Although some projects will require the delivery of the whole system at the same time, a versioned approach is easier on all parties, as the partitioning reduces the complexity of each phase. It also reduces the risk of being completely wrong or of tying your system into obsolete technology.

Also, somewhere along the line, conversion costs must be taken into consideration when producing the Cost-Benefit Statement. Some of that mysterious "costing data" coming into the process must be what is known about conversion. This early in the development process it will be very difficult to estimate the actual cost of data conversion, since the system hasn't been designed yet and all the details of its implementation aren't in. But the client still wants to know something about the cost. So you make the best guess you can with the knowledge at hand. As more data becomes available it will affect the cost-benefit estimates. The immediate impact of conversion on development costs can only be lessened by spreading the need to convert across several versions of the system to be developed. Again, this reduces the risk to the client.

For each proposed version of the system, the Cost-Benefit Statement should contain four alternatives:

1. *Do nothing* (or maintain the status quo), the base against which the other options will be measured;
2. Act on the identified *"immediate opportunities for improvement"* (IOIs);
3. *Develop* a new system;
4. *Buy* a software package.

"Do Nothing" Option

Do nothing really means keep right on doing whatever has been done all along up until now. In other words, the evaluation of the new Essential Business Model suggested that the system currently in place is about as good as can be expected at this time. Hardly any team will ever arrive at this conclusion, if for no other reason than it simply isn't done! Sadly, it's also traditional to recommend that everything be automated. This usually happens because the Development Team typically only models computer systems rather than business systems.

Another reason why *do nothing* is an alternative that must be fully costed is that it reflects today's reality; this is the only way to measure the value of other alternatives, since we must measure *against* something! The cost of doing nothing is exactly the same as the cost of the current system; since we have to know the cost of the current system to measure benefit, it's easy to include it as a costed alternative. Besides, it is a valid alternative to the client, and it's important to be able to see a specific cost for the status quo.

Be careful not to fall into the trap of thinking that the *do nothing* alternative is free. As well as normal running costs for any portion of the system currently automated, the current cost of all manual processes must also be included, because in a new implementation the automation boundary may shift. Also, the cost of any outstanding maintenance items that cannot be avoided (and that would presumably get picked up in any major system reconstruction) must be included in the *do nothing* alternative, as well as the cost of the project so far.

"Immediate Opportunities for Improvement" Option

An IOI is an "immediate opportunity for improvement." These critters are often discovered during the creation (or revision) and evaluation of an

Essential Business Model, which contains models of the substance of both data and process. The IOIs are simple, everyday things like replacing a report with a series of screens, changing to a less expensive long distance carrier, putting an additional door in the wall between the stock room and order department, or correcting an edit procedure to prevent bad data from getting into the system. These are productive little jobs that can be undertaken almost immediately, finished quickly, with a substantial return on the investment. These IOIs are generally "discovered" because most mature companies have older systems with inconsistent or incomplete documentation. To produce an Essential Business Model often means that a *current physical* systems model will be done first. If this is the case, all kinds of possible improvements are likely to be discovered. Quite often there is the opportunity to correct immediately the source of an old nagging problem in a system, either by the data processing staff or by the user, with the resulting positive effect. These opportunities are usually inexpensive and easy to execute and sometimes don't even involve fixing the code.

Because IOIs are usually discovered while doing the kind of archeology that creates the Essential Business Model, they should be part of the possible solution for the user. Sometimes, repairing IOIs can be the best solution to a business problem, and therefore should be included as one of the options to be evaluated in the Cost-Benefit Statement.

IOIs are always done to the current, in-place system.

"Develop a New System" Option

Develop a new system means estimating the delivery of the Functional System Model, design, construction, implementation, training, conversion, and all the other things that go into building the system. Because you have an Essential Model, estimating a custom-made system will be somewhat easier than it might normally be.

Buy a Software Package Option

You no doubt noticed that physical alternatives, in the conventional sense, were not extensively discussed as part of the Cost-Benefit Statement. That's because automation boundaries, selection of processors, what processes merit batch or online design—all such physical things—have only been guessed at with the first iteration of the Essential Business Model. These educated guesses have been made so that corporate management will have a feel for the operational feasibility of the business system. However, at this point, not only can the answers to real physical alternatives not be provided, but some of the questions aren't yet clear. Serious discussions of physical alternatives are only appropriate after the Functional System Model has been delivered.

The Essential Business Model represents what must be done to meet the needs of the user, but does not specify how, when, with what technology, or how it will be controlled. In other words, it is completely implementation-*independent*. By virtue of the name alone, "physical alternatives" are totally implementation-*dependent*. At this time, then, it's folly to select processors for processes that have yet to be fully defined. It is difficult to state what will be online when definitions of data, relationships, and what needs to be done with the data in order to implement the system are less than clearly understood by the users or the Development Team, although there may be confidence in opinions. It's equally difficult to recommend that packaged software can do the job.

Even though it's common practice and an industry-wide malaise to select physical implementations—and doing cost-benefits on that basis—before even knowing how the data classes, relationships, and processes specifically go together, we feel a substantially different approach should be taken.

To minimize the risk of building technologically inflexible systems, we believe decisions on automation boundaries and processors must wait until some fairly detailed version of the Process Model has been done. To this end, the Cost-Benefit Statement should use *do nothing* as the base, and *IOIs* and *build a new system* as the only other options. Buying a software package should not be an alternative at this point. However, reality often dictates that purchase of packaged software be considered right at the start of the project. We don't subscribe to this notion, nor do we feel it's very reasonable, but it certainly is fashionable, so we have provided for its inclusion in the Cost-Benefit Statement prepared with the Essential Business Model. We will discuss this option in more detail in the next section, when it becomes an honest reality for the developers.

The *develop a new system* option doesn't specify the new system; it's more like looking at the shadow of the new system, which is still lurking around the corner. It's a place-holder, which marks one of the four major options. If the client elects to proceed with this alternative, the only real commitment is to some more analysis and specification. Real technology should only be added after all the functions have been specified (which we'll review in the next section). Granted, this will cause a change to the cost-benefit somewhat later in the development process, but that's reality too.

As we said earlier, it's practically impossible to cost-benefit and assess operational feasibility without making some technological assumptions. That's why the Development Team, with the client, will make some guesses—but not decisions. It will probably be assumed that certain parts of the system will be automated, even online and interactive, and that some parts of the system are micro-based. These are probably safe guesses and can be accomplished by "blitzing" the first version of the Processor Configuration Model. There should be a lot of knowledge and experience available between the Development Team (TD) and the client. But the project manager must ensure that technological assumptions do not prejudice the implementation of the Process Model.

Sec. 4.2   The Detailed Diagrams   109

Any system in which complete, up-to-date information must be available on demand (such as monitoring automated production systems, financial markets, airline reservation systems, or the weather) is a reasonable candidate for some online functions. An environment in which people work on unrelated segments of confidential data (such as accountants or income tax firms) might use some form of microcomputer support. A business that is going to have to respond to customer phone calls will need a switchboard and a supply of canned music! So what happens if the implementation technology changes substantially from what was assumed the first time around?

The impact on the Cost-Benefit Statement after technology has been added to the systems model is often not very significant. Assuming the Development Team, who made the assumptions in the first place, are top-caliber people, you'll find they usually aren't very wrong. Their experience will tell them what's reasonable. But be prepared, just in case. It's even possible the payback to the client will increase from the original benefits after applying the new technology. But if payback decreases, it may be wise to spend more time on versioning (and preparing a new Version Plan) with the client.

Consolidating the Essential Models

The final process in creating an Essential Business Model is really not much more than a packaging affair. See Fig. 55.

The three major deliverables should now be brought together nicely to make a cohesive "package." Only one activity is really outstanding in this process. Both the Essential Process Model and the Essential Information Model had separate data dictionaries (perhaps). Since the Information Model should contain the knowledge of the system, and the Process Model should contain what to do with that knowledge, it seems reasonable enough to have only one data dictionary. Also, the same data dictionary should contain all the text, including relationship definitions, from the Information Model. If there's anything to do here, then, it is to combine the two data dictionaries. With an automated data dictionary, their integration is not a problem.

Not everyone agrees with the idea of putting relationship and data element definitions in a single data dictionary, since they have different formats. So be it. A regular dictionary doesn't have exactly the same format throughout either. We have found a combined data dictionary to be very useful, since the rules have to be looked up *somewhere* and this eliminates the need for a second dictionary. We've come to the conclusion that there should only be one data dictionary rather than two. Having only one makes life somewhat duller, but it certainly makes it more comfortable!

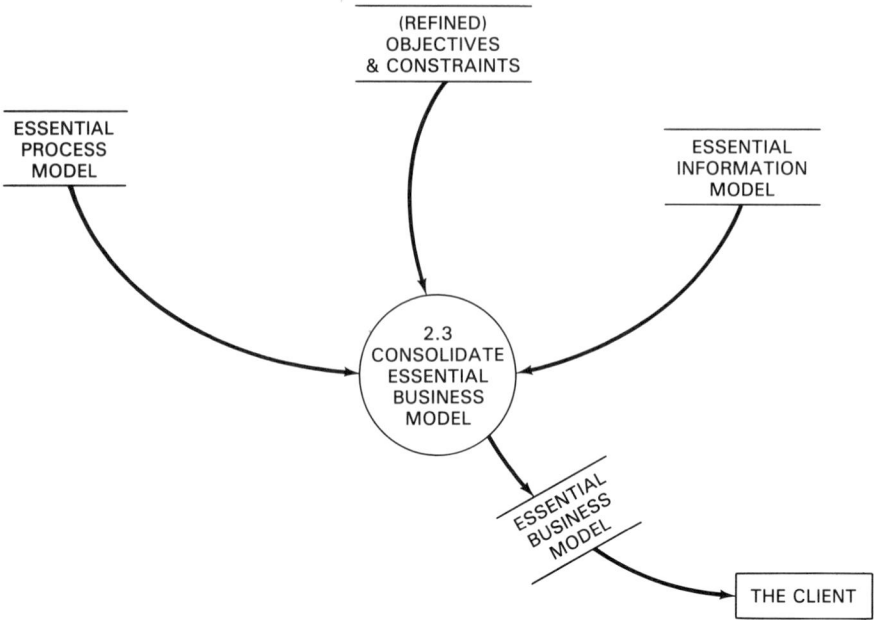

Figure 55: Consolidate the Essential Business Model (2.3)

Essential Business Models and Strategic Planning

As your organization implements Essential Business Models of various applications, departments, or business patterns, there will eventually be a desire to concatenate, or bring together, these EBMs to form what we have come to call the EBM SuperModel.[10]

Each system developed, such as an inventory management system for a microcomputer retail outlet, will have a context level diagram associated with its Essential Process Model. The high-level model in Fig. 56 illustrates the net inputs and outputs of the system's basic inventory management process. Each interface with the system, illustrated by a box, is a terminator representing a person (and functional responsibility), a system, or another department (which is really another system). Each of the terminators could and should have its own Essential Process Model, unless the terminator is outside the company, such as the bank, a government tax department, or perhaps a client of the company.

---

[10] A *supersystem* is defined in *Webster's* as "a system that is made up of systems." A SuperModel, therefore, is a high-level business systems model that is joined with other high-level business systems models to make a whole. With respect to data models, James Martin calls this "canonical synthesis," that is, integration of third-normal form data models by application into an organizational model, independent of the implementation technology or individual applications of the data; an "essential" data model (*Managing the Data-Base Environment*, Englewood, Cliffs, N.J.: Prentice-Hall, 1983). pp. 235–276.

Sec. 4.2   The Detailed Diagrams   111

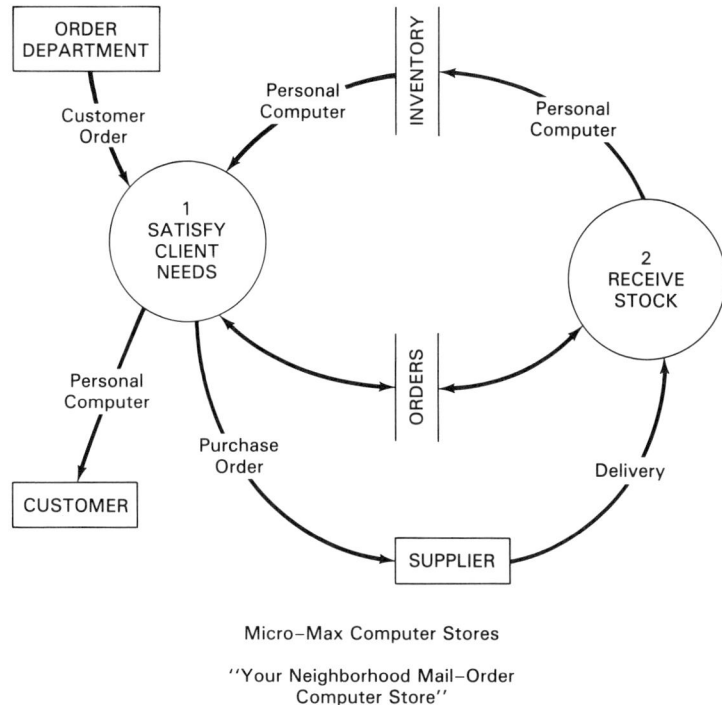

Figure 56: EPM for an Inventory Management System

Once Essential Process and Essential Information models have been developed for several applications, including the interfacing terminators on those models, they can be concatenated to form a SuperModel of the business your company is in, or at least part of that business (see Fig. 57). This can give you a functional architecture of the business and, essentially, how it operates and how the data is handled and shared.

These SuperModels will eventually lead to an improved understanding of the complexity and interrelatedness of the company's processes and business patterns. They will provide business analysts with the opportunity to ask rational, knowledgeable questions based on larger, company-wide models. They will also be able to better assess exponentially positive or negative effects across interrelated systems when planning the development of support functions for new products or services by the company.

This knowledge of data and processes and what can be done with them can help to deliver the support systems much faster than usual when the company plans to introduce a new product or service to the marketplace. A complete blueprint of the essential processes and data of the business can also prevent poor or untimely business decisions.

The Essential Business Model is truly the innermost substance of the

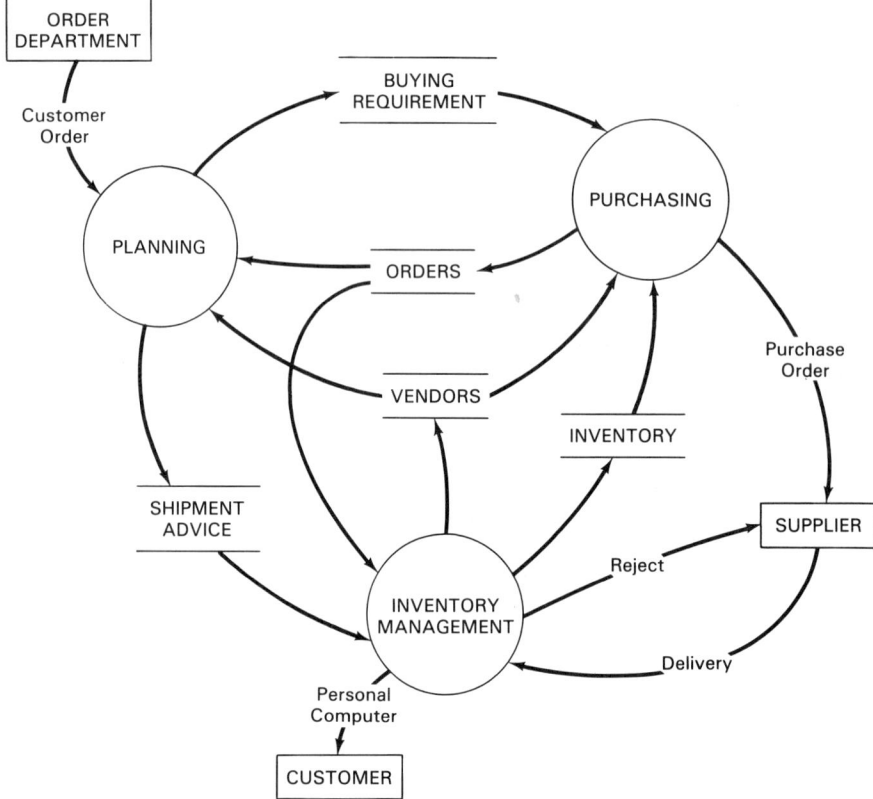

Figure 57: An EPM SuperModel of Three Departments

business. Accordingly, systems developers and users might take great pains to ensure that they get that substance right.

4.2.3 Model the System Functions

The Functional System Model, like the Essential Business Model, contains both an Information Model and a Process Model. The Functional Model, though, is a *completely detailed specification* of the user's system, far more complete and rigorous than its earlier business version. It documents all the data and all the transformations of data that will exist in the new system.

It's quite important at this point to ensure that both the Information Model and the Process Model are complete, so that the client and the Development Team can verify that all the things that have to be done by the

Sec. 4.2   The Detailed Diagrams   113

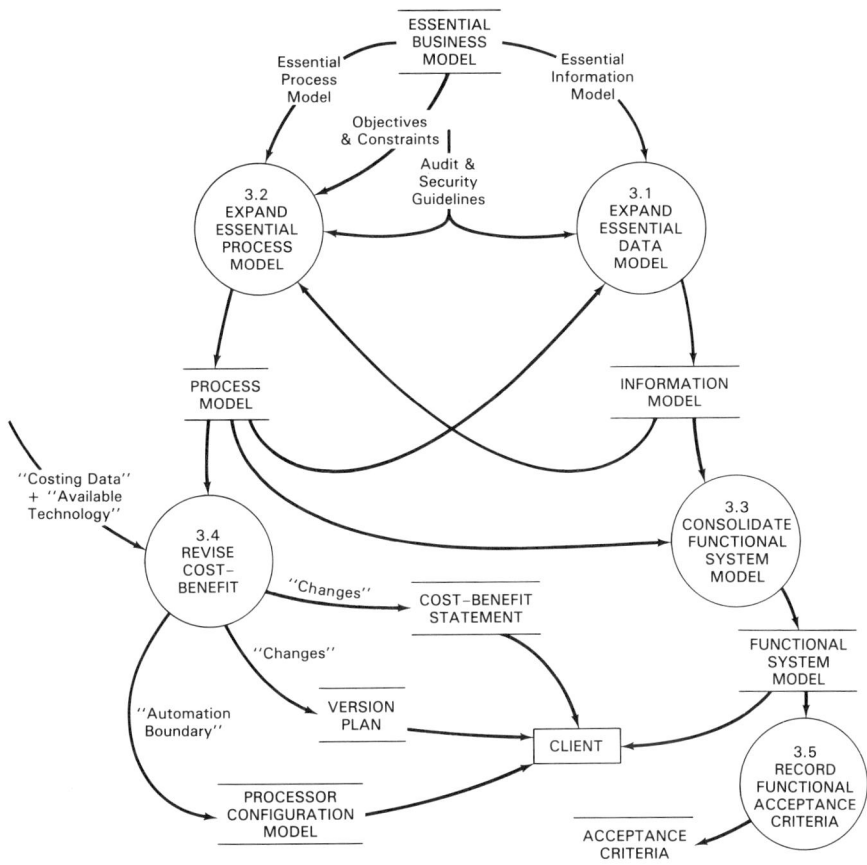

Figure 58:   Model System Functions (3)

system are being done and that all the data to do it are available. The Process Model "can be thought of as an output producing machine [and] a mechanical error in such a model involves an inability to produce the desired output from the available input." The Information Model "is a question answering machine; a mechanical error in this type of model shows up as the inability to answer a desired question."[11] In other words, the *what* must be complete. We deal with the *how* (or processor-related details) in design.

The diagrams of the lower level processes in Fig. 58 are very similar to the breakdown of activity 2, *Create Essential Business Model*, discussed earlier (Fig. 21). The Functional System Model is simply an *expanded* version of the Essential Business Model, with much greater detail in its component parts.

[11] Paul T. Ward, *Systems Development Without Pain* (New York: Yourdon Press, 1984), p. 85.

However, the Information Model is no longer pure. It now expands the Essential Information Model into data groups (i.e., "objects") and how they relate to each other in the real world. The complete Information Model contains all the information—an entity-relationship diagram, textual specifications for each "object," the data elements attributed to each "object," and textual specifications for each relationship—to support the proposed system. Policies and conventions such as edits, audits, and approvals are a large part of this model of the client's system.

We consider the delivery of a good Information Model as a prerequisite to delivering good data inventory. Without a quality business database (i.e., clear identification of business entities and policies), it will be extremely difficult to identify and capture essential corporate and application data.

It is a fundamental premise that before business patterns can have any real meaning, the rules, conventions, and laws that govern the operation of the business must be understood and defined. To achieve this, the Information Model identifies data groups and how those groups (or elements) relate to each other in the real, imperfect world.

Although the Information and Process models again "feed" off each other, the Information Model significantly influences the Process Model. The text component of the Information Model specifies all the rules of the game—that is, the rules, policies, laws, and conventions behind the relationships among data groups. The Information Model, however, doesn't specify exactly how these "rules" should be implemented. That's the job of the Process Model. But the Information Model, in terms of company policy, must be totally predictive when it is complete; that is, it must be able to answer any question concerning what the system must know to be able to function.

An accurate and complete Information Model, once translated into a physical database, can provide the user and the company with an excellent data inventory. Without the right data it is impossible to do things for users who urgently need to get things done! Without the Information Model, the "things done" to and with data will be haphazard and certainly not based on knowledge.

We cannot overstress that data and their custody must be treated as a corporate resource and responsibility, rather than some incidental stuff that passes through the system to produce today's reporting needs.

Figure 59 is an example of the entity-relationship diagram component of an Information Model for a human resources system; Fig. 60 illustrates a "relationship" definition; and Fig. 61 shows a data dictionary entry for one of its "objects."

There are clearly more detailed methods of recording the preceding data. Data administration types will no doubt be more rigorous in their approach. But for analysts, our philosophy is "whatever works for you, do it," as long as you do not *not* do it!

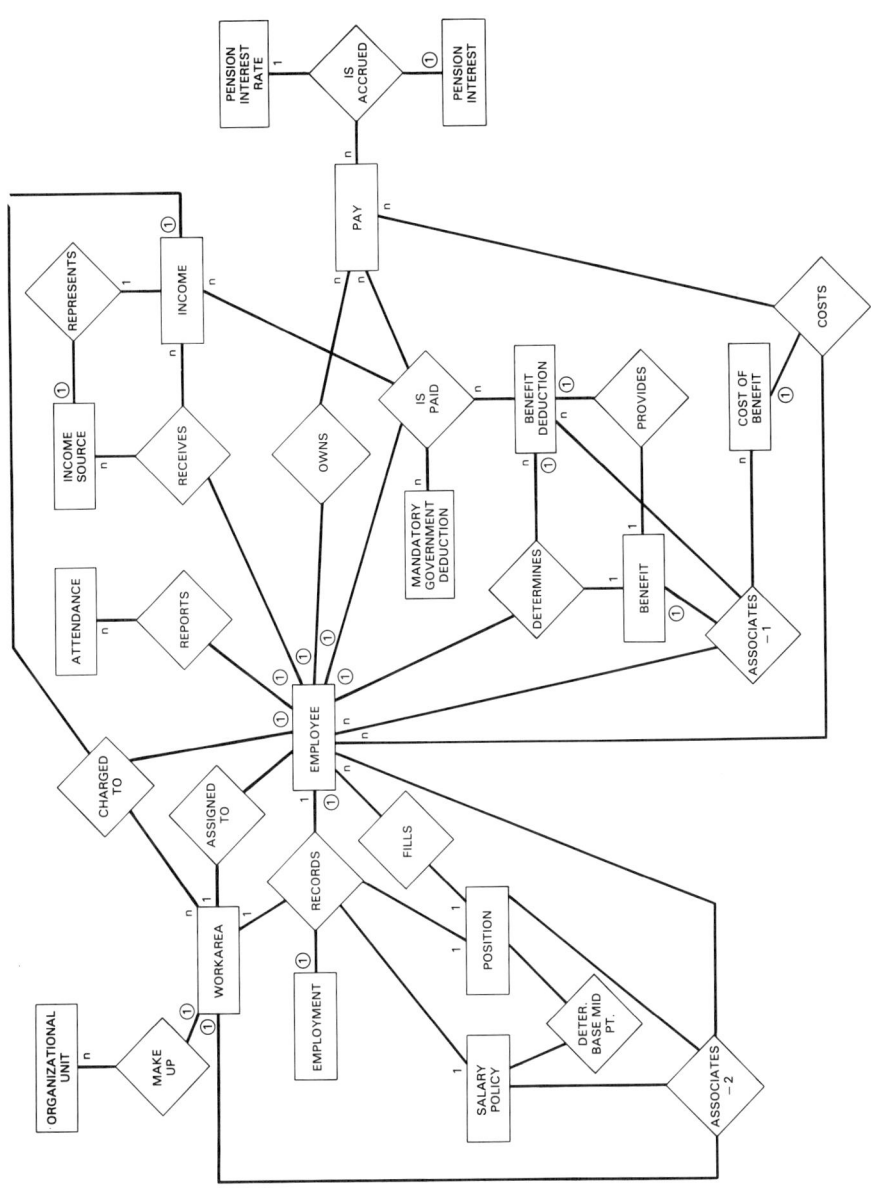

Figure 59: Entity-Relationship Diagram of a Human Resources System

116   The Substance of the Model   Chap. 4

The expanded Process Model also contains all sorts of control functions, specifically the execution of the edits, audits, and approvals required by the system.[12]

RECORDS

For each EMPLOYMENT there is an EMPLOYEE, a WORK AREA, a POSITION, and a SALARY POLICY

associated

if the EMPLOYMENT records the effective date of the EMPLOYEE's career change;

such as: commencement
       rehire
       work area change
       position (job) change
       termination
       status change
       salary change
       job points change

and

if the EMPLOYMENT states the POSITION, WORK AREA, and compa-ratio (derived from SALARY POLICY) assigned to the EMPLOYEE.

Reverse cardinality.

For each EMPLOYEE there can be one or more occurrences of EMPLOYMENT, but only one occurrence of WORK AREA, POSITION and SALARY POLICY.

Identifier • Job number
          • Work area number
          • Midpoint salary (base salary + (slope * total job points) * (job differential/100) * (regional differential/100))

Figure 60:   Data Dictionary Entry for Relationship "Records"

[12]Allan Brill's *Building Controls into Structured Systems* (New York: Yourdon Press, 1983) deals with the controls issue perhaps more thoroughly than any other recent book and discusses how controls are illustrated in process models.

Sec. 4.2   The Detailed Diagrams   **117**

Figure 61: Data Dictionary Entry for Object "Employee"

EMPLOYEE

- An employee, excluding agents and branch managers.
- Including—People who have been hired but have not started work
    —Current employees working
    —Current employees on leave
    —Any employee who has terminated, retired, died
- Can be located in the U.S., Canada, or the International group.
- There is a requirement for a minimum amount of data on an employee to be kept forever.

Identifier—Employee number

- Cross reference employee number
- SSN/SIN
- Name
- Address
- Home phone number
- Language
- Pay data—Pay currency
    —Pay method
    —Pay frequency
- Number of dependents
- Work hours
- Current status—Work arrangement
    —Work availability
    —Staff status
- Tax exemption details—multivalued (state, federal, municipal)
- Hire method
- Education data—multivalued
- Previous experience

Employment date

- Net credited service date
- Pension—Pension class
    —Pension date
    —Pension beneficiary name
    —Pension beneficiary relationship

- Insurance—GI beneficiary name
  —GI beneficiary relationship
  —GI assignment flag
  —Guaranteed GI amount
  —Smoker/nonsmoker
  —Dental coverage option
  —CMP coverage option
  —HMO coverage
  —Workers compensation coverage
  —Waiver (multivalued)
- Bank—Bank number
  —Bank transit number
  —Account type
  —Account number
  —Account 2 type
  —Account 2 number
  —Account 2 amount
- EEOC code
- Handicapped
- Visible minority
- Spouse—Spouse name
  —Spouse Social Security number
  —Spouse birth date
  —Spouse language
- Living residence code
- Work residence code
- Work address
- Work phone number
- Space allocation
- Salary review—Date
  —Frequency
- Performance review—Date
  —Rating
  —Acceptances
- Preferred first name
- Sex
- Security signal
- Marital status
- Supervisor employee number
- Building code

Sec. 4.2 The Detailed Diagrams   **119**

- Birth date
- Termination—Termination code
  —Termination date
  —Rehire eligibility
- Sick days credit (derived from attendance)
- Vacation days (derived from attendance)
- Miscellaneous text

Data Dependencies.

1. If there are no ATTENDANCE occurrences for an employee, this indicates that employee has not taken any vacation or sick leave.
2. For each ATTENDANCE, EMPLOYMENT, INCOME, BENEFIT DEDUCTION, PAY, and PENSION INTEREST occurrences there must be a corresponding EMPLOYEE.
3. If there are no EMPLOYEEs, there will be no COST OF BENEFIT

Note: It is currently possible for an employee to have more than one employee number. A new (different) employee number is assigned:

- If the employee retires and continues/restarts work
- If the employee transfers territory
- If the employee restarts work with no vested pension from previous company employment

In the above cases, multiple employee numbers have been a requirement for Benefit Plans purposes; this conflicts with the requirement for a single employee number for an employee for tracking pay and career and for building and computer security purposes. A policy decision is required on this matter before design of a new HR system.

Audit and Security

In Fig. 58 notice that **Audit & Security Guidelines** is input to the processes that expand the Essential Information Model and the Essential-Process Model. An extract from that diagram is shown in Fig. 62. This information, and what it causes the system developers to do to the Essential models, is, next to the user's basic requirements, perhaps the most important and time-consuming aspect of developing the Functional System Model. So, where did these mysterious guidelines come from?

Almost every company has some guidelines for systems developers related to security and audit issues. Some are more demanding than others.

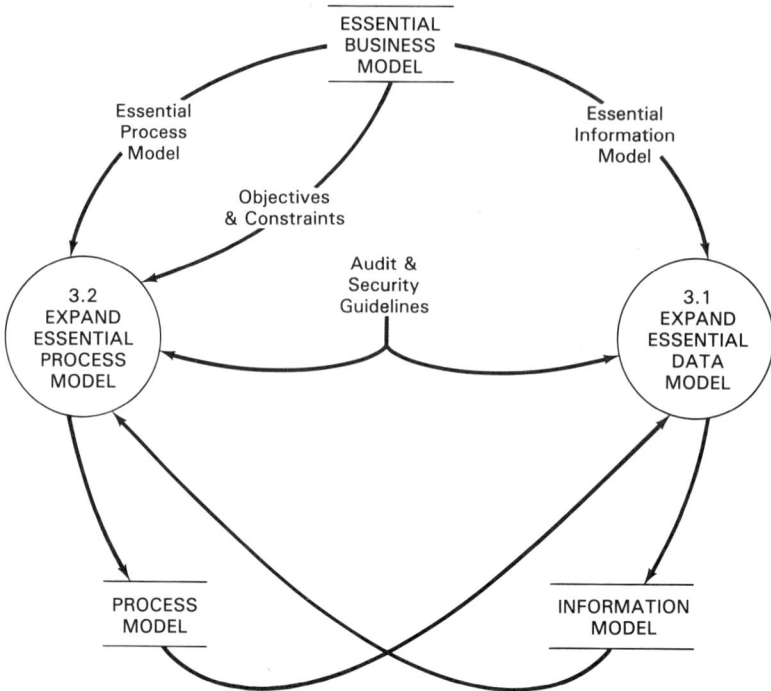

Figure 62: Expanding the Essential Models with Physical Control Functions

These guidelines should identify to the Development Team general audit, security, processing, recovery, and environmental issues that are (or should be) of concern to audit and security personnel. It should also be of great concern to the systems developers since these functions make up the major part of the work that goes into building a strong and cohesive system. In the Dictionary entry for this data flow we have identified a number of issues that should be considered by the Development Team when preparing the components of the Functional System Model.

It has been suggested to us that the only thing we would think of to say about audit and security in the context of the SuperSet model—without choking—would be that the use of the SuperSet forces (reminds?) us that audit and security requirements should be considered early in the development process. And work done early (or up front) is work done more cheaply.

Well, we want to say more than that—and without choking. We certainly want to make the point that, although audit, security, and control components of the delivered system are implementation details, and not part of the Essential Business Model (i.e., not related to fundamental business activities), the definition of audit and control features is nevertheless an essential activity of the development process.

Although we have identified the need to address audit and security requirements throughout the systems development process, it has also been our experience that people are often unclear about how systems security, controllers, and internal audit requirements and involvement differ. All three may be talking control. So let's clarify.

When application systems are being developed, whether for the mainframe or some other computer system, there are always control issues that must be addressed. These issues may include, depending on the application and computer environment, physical security of the hardware and software, recovery capabilities, data security and privacy, organizational controls, accounting controls, controls to ensure operational efficiency and effectiveness, audit and management trails, and various other application controls to ensure that processing and record keeping are complete, accurate, and authorized.

Not surprisingly, different groups represent different interests. The separation of controls is not simply black and white. But, in general:

- *DP Security* is concerned with physical security, data security, and recovery abilities in general. But their interest is more specifically in the area of mainframe environmental controls and their impact on applications (e.g., disaster recovery and computer security). They are concerned that applications apply the available tools (e.g., ACF2) and adhere to departmental standards (e.g., backup and disaster planning). DP security may, therefore, have certain security and recovery requirements that should be considered when developing a system. These standard elements are included in Audit & Security Guidelines.

    DP security should also be a consulting source on security and recovery issues. They should also be available to assist in the design and testing of security and recovery capabilities.

- *Controllers*, on the other hand, are concerned that data being passed to the corporate accounting database is complete, accurate, and auditable. Controllers, therefore, may have some key accounting control requirements (e.g., general ledger/subledger reconciliation or suspense reconciliation procedures). In most companies, controllers are also a consulting source on accounting controls, and they may be able to assist in the design and testing of these accounting controls.

- *Internal Audit* is concerned with application controls of all kinds, whether they are security, accounting, or any other type. Internal audit can't design solutions without the risk of losing their assessment objectivity, but they can assist in the requirements definition, that is, many of the control components that have to be built into the Process Model segment of the Functional System Model. In any company, internal audit should also be available as a consulting source. They should also attend walkthroughs at which they can assess the adequacy of controls in all areas. Normal application control requirements, covering a broad range, are also spelled out in Audit & Security Guidelines.

Another area of concern when building systems today has to do with the threat from within the organization.

The incredible recent and rapid increase in computer literacy, coupled with the undeniable proliferation of micros ("This thing could eat you up before we even know it's hungry!"), presents entirely new challenges to the system architect, security specialists, and the EDP auditor. Contrary to perceptions popularized by the news media, these challenges are not just from the external *hacker*. Potential threats are also from past and present employees.

Agreed, a hacker, depending on her or his level of expertise, could penetrate a computer system by using a modem and programming a micro to guess various passwords. Once the password has been compromised, access can be gained to information. However, most security systems (packages like ACF2, RACF, SECURE, TOPSECRET, or SAC) limit the information that a user-ID can legitimately access. For example, if a logon-ID/password is compromised that allows a person to read only nonsensitive files, then the threat is minimal. Conversely, if a more sensitive logon-ID/password is compromised, the exposure is increased in direct proportion to the access authority. This is where application controls such as those identified in Audit & Security Guidelines come into play.

Since the casual hacker doesn't usually have knowledge of the business functions that ensure that authenticity of the access, well-constructed controls should pick up erroneous data. But if the hacker is a past employee who has a working knowledge of the business functions and application controls built into the system, it would be much more difficult to trace.

Possibly the greatest threat to businesses and systems are current employees, because they often have authorized access to files (both automated and manual) and they often have a working knowledge of the application controls. While most employees are honest, the increased threat from past and present employees is very real.

Although systems development personnel may well be trained in systems analysis and design, they are rarely trained well in the esoteric design of controls—even if surrounding essential processes with physical controls is the primary purpose of the Functional System Model! On the other hand, systems security officers and EDP auditors (or "control experts") *require* systems developers to apply controls that they think a good system should contain, in order to protect against threats perceived or real. The rub is that the two sides hardly ever talk to each other; and systems developers continue to make the same mistakes from one system to the next, while audit and security specialists continue to write meaningless reports after all the damage has been done—to no avail.

Sec. 4.2 The Detailed Diagrams 123

Problems in systems don't simply materialize—they evolve. A virtual plethora of exposures, errors and omissions are built into almost every system. The answer, of course, is to:

- prevent as many of them as possible from occurring;
- detect them, should they occur;
- correct them as soon as possible.

The solution to this old problem is for the systems developers and the security and audit specialists to work together during the development process, rather than imposing sporadic controls on the system as an afterthought. In other words, EDP auditors and systems security analysts must be legitimate part-time advisers to the Development Team.

Unless this happens—and it's especially tough to get EDP auditors to agree—project managers and systems developers will continue to view EDP auditors as those "who sit on the hillside observing the battle, and after it is over, shoot the wounded."

It's crucial that specific requirements of auditors or controllers be included in the specifications of the Functional System Model. Audit should be involved as early as possible and control functions engineered into the system, rather than integrated after the fact. If auditors are involved in the detailed function modeling process they will be able to point out danger areas, the "open doors and windows" of the system. But we agree that they shouldn't dictate solutions; that would constitute a conflict of interest.

Once all the essential functional controls—edits, audits, approvals—have been specified for the system, it still isn't quite *physical*. The model of the system still needs to have processors, automation boundaries, and substantial design applied to it—and then even more controls added—but it's getting close. We might now call it *physiological*, being a bit of both. To illustrate its physiological nature, this model must include all physical constraints that are known to the Development Team and are outside their ability to change. This might include premature technological decisions (the boss insists on the DataSchwartz 2938, which only has a card reader!), imperfect or redundant methods of data transportation, external constraints (such as specific methods of collecting data for another system), or antiquated technology imposed by some other interface. The Process Model must specify all system-planned responses, with the stimuli and data essential to the processes that deliver the planned responses.

Figure 63 is an example of part of a Process Model from the same human resources system shown earlier. Figure 64 is a corresponding data dictionary entry defining a data flow; and Fig. 65 illustrates a process specification written in plain, ordinary, user-friendly English.

The Process Model and the Information Model must be totally predictive when they are complete; that is, they must be able to answer any question

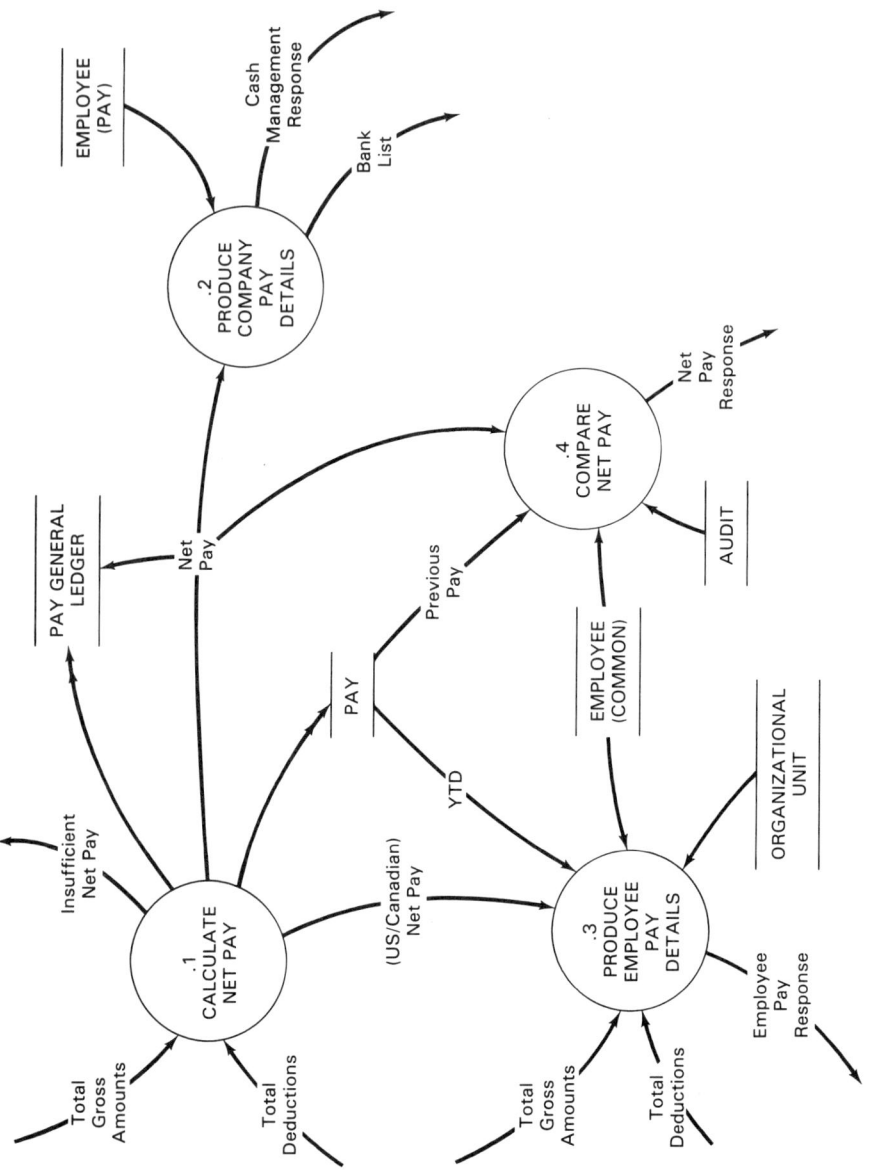

Figure 63: Sample Data Flow Diagram from a Human Resources System

Sec. 4.2 The Detailed Diagrams

```
INSUFFICIENT NET PAY
* advice to Payroll of deductions not taken from employees due
  to insufficient net pay
={EMPLOYEE NUMBER + NAME + {[BENEFIT NAME;
  BENEFIT CODE] + DEDUCTION AMOUNT}}
```

Figure 64: Data Dictionary Entry—Insufficient Net Pay

concerning what the system must know to be able to function and be able to answer any question concerning the system's behavior in a specific environment.

Distinct from the Essential models, the Functional System Model is *not* usually very simple when completed.

The Cost-Benefit Statement Revisited

Once again, when we look inside process 3.4, *Revise Cost-Benefit* (Fig. 66), we find that we can now almost finalize the automation boundary by using the Process Model. We can then see how this affects our Version

```
CALCULATE NET PAY: process 1.2.4.4.3.1

Subtract TOTAL DEDUCTIONS from TOTAL GROSS AMOUNTS to get
NET PAY

    if NET PAY is less than zero
        Find all PAY AMOUNTS for deductions taken this pay from PAY;
        Determine PRIORITY FLAG from BENEFIT DEDUCTION using PAY
        CODE;
        Subtract PAY AMOUNTS for the lowest priority
        BENEFIT DEDUCTIONS from TOTAL DEDUCTIONS until
        NET PAY is greater or equal to zero;
        For each PAY AMOUNT subtracted
            delete PAY
            delete PAY GENERAL LEDGER;
        Issue INSUFFICIENT NET PAY

create PAY GENERAL LEDGER

issue NET PAY
```

Figure 65: Sample Plain-Language Process Specification

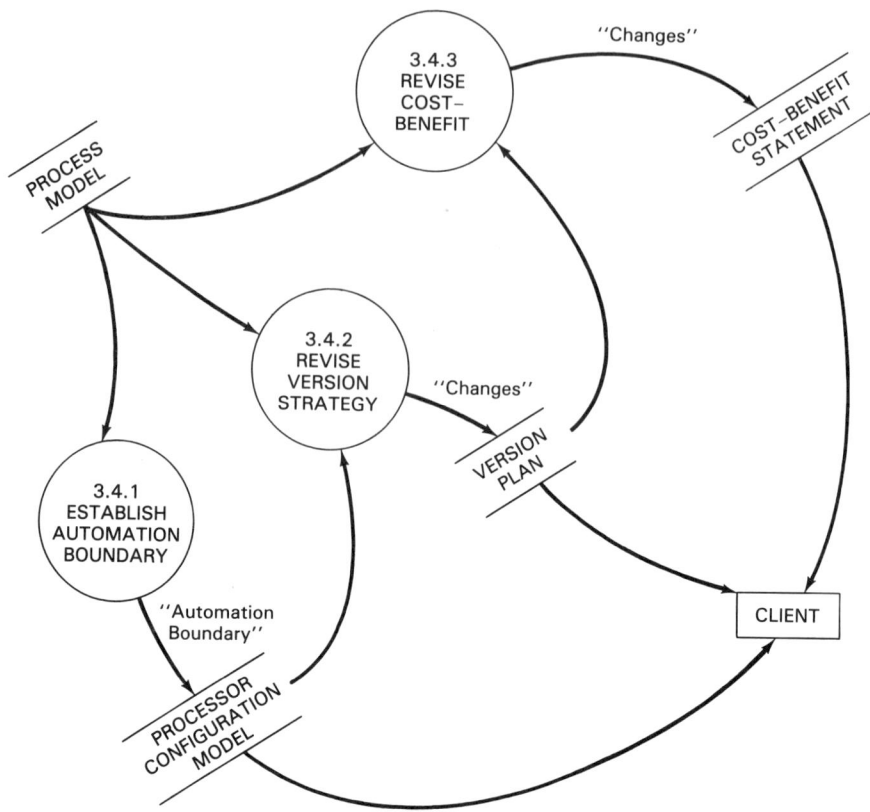

Figure 66: Revise Cost-Benefit (3.4)

Plan—the one produced earlier for the client—and how that in turn will affect the Cost-Benefit Statement. Although the person/machine boundary has been determined, automated processors have yet to be finalized, since processor selection will be greatly influenced by design considerations. While the Processor Configuration Model will be completed in Design, its final form will invariably cause a change to the person/machine boundary, thus again causing more amendments to the person/machine boundary of the system. But in order to prepare a revised Cost-Benefit Statement for the client, probable boundaries can be determined at this time. Certainly throughout this process there will be a growing awareness of which functions will probably be automated and which won't, and it's foolish to pretend ignorance (especially when pretense isn't necessary in so many other areas). Client constraints may also dictate from the very beginning of the project that certain functions will be automated. Like Dijkstra, we will follow the rules 90 percent of the time, but always keep a look out for the little red lamps flashing in the corner of our eyes.

Most organizations do cost-benefit analysis rather poorly. This comes as no surprise to most of us, since it's usually very difficult to establish the cost of something when we're not entirely sure what that something will be. For the same reason, it's difficult to calculate the benefit. And quite often, the cost or benefit of a system is directly associated with *opportunity* gained or lost. And who's to say (usually, anyway) what the dollar value of a specific opportunity may be?

There may be instances when opportunity can be quantified. For example, if a system can be implemented that will give a more accurate reflection of current cash flow, check float and receivables, that system's precision and timeliness of data could allow company money managers to utilize more fully current cash positions in the overnight investment market. If an additional $1,000,000 is "freed up" for investment, this would translate to a benefit of about $610,000 over five years (at a rate of 10 percent per year). This is a direct opportunity gained, that is, a direct benefit received as a result of developing a new and more accurate system.

Another example of a change producing a quantifiable benefit might be that of a company implementing a system using electronic funds transfer to pay its employees and major creditors. This system, rather than producing greater revenue as in the first example, justifies itself by reducing operating costs.

The other side of the opportunity-benefit equation is the opportunity lost. In the first example, *revenue would be lower by $610,000 after five years if the system had not been built.* It's therefore crucial that current operating costs and revenues be fully expressed in the *do nothing* option of the Cost-Benefit Statement, so that all changes to these figures will clearly stand out as a result of the creation of any new system.

This applies equally well when comparing types of implementations, such as online compared to batch processing. It's always possible to quantify opportunity, but it is also difficult and requires that the user agree with the numbers.

For example, for a retail outlet (let's say a shoe store) to remain competitive, it must accept major credit cards as well as cash. It's commonly accepted that the shoe store will lose customers (that is, they will buy their shoes somewhere else) if it does not accept major credit cards. The question is, How many customers will the store lose? How much money will that translate into? Won't the customers come back with cash anyway when they have decided what they want? Will it matter? The accepted wisdom is that it matters greatly. It's also argued that some customers will intentionally stay away from the store just because it doesn't accept payment by credit card. But how many customers and how much money will the store never see because it does not provide a facility such customers want?

The same question is faced by banks when deciding whether to provide online, multibranch banking to customers; and to insurance companies when questioning the need to provide more timely (and more expensive) quotation systems to agents. Each company, in its own way, must address the issue of *value* of the opportunity; and each person's perception of the value will differ. For this reason, cost-benefit analysis must no longer only be done by DP staffers on a project. Certainly, these numbers can be provided by the users or a marketing research group.

User analysts must prepare the Cost-Benefit Statement with the DP systems analyst acting as an adviser and consultant, because it's the user's perception of benefit that counts in the end. Also, only the user knows exactly what the costs and benefits are in the part of the system that is outside the automation boundary.

Also, if senior users aren't directly involved, there is always the danger that DP members of the Development Team will understate the cost and overstate the benefits, because it may be in their interest to do the work—either because they need to find the work, or because it's an interesting high-tech application, or because the selected approach is their personal pet solution.

The user's approval of the third option we have suggested in this book, *design a new system*, should not mean build a new system in-house. It should simply mean deliver a new Functional System Model. And after the new Process Model and detailed Information Model have been delivered, another revised Cost-Benefit Statement should be prepared, with the appropriate recommendations. Then, if the user likes it, the Development Team may get approval either to build a new system in-house or to evaluate and possibly purchase a software package.

But when the Functional System Model is delivered, it is very correct to discuss real physical alternatives. Until this time, however, it is ludicrous to attempt to allocate processes to processors that have yet to be defined. It's difficult to state which processes will be online and which ones will be batch when definitions of data, rules, and what to do with the data are less than perfectly known. There may be strongly held opinions, but these are not facts.

The Development Team, with the concurrence of the client, must and will make some technological assumptions, such as tentative automation boundaries, but they will not make decisions. (In some companies users often perceive the technological assumptions to be promises, and the costs final. Small wonder that estimates are often 100 percent to 500 percent off!) The team will probably even assume parts of the system to be automated, and perhaps some of it to be online. These are usually safe assumptions, since some thinking has gone into the reasons why the project exists in the first place. With reasonable implementation assumptions a Cost-Benefit Statement can be delivered to the client that covers the *design a new system* option.

In the end, there should be at least two Cost-Benefit Statements delivered: the first one with the Essential Business Model; the revised version with the Functional System Model.

If, on the basis of the second Cost-Benefit Statement, which covers the three standard options; the *design a new system* option is chosen, then a third Cost-Benefit Statement must be done that itemizes only the alternative implementations and strategies of the new system—for example, (1) batch implementation only; (2) online, real-time to the user (i.e., using shadow files); (3) online *real* real-time; (4) purchased software packages; (5) contracted design and programming; (6) in-house development; (7) contract design, with in-house programming and implementation; and so on. The *do nothing* option is still an alternative; the user can always pull the plug.

This approach to preparing the Cost-Benefit Statement—making sure the user does it, and making sure there are several iterations—will not only free up some time for the DP systems analysts to do some less political things, but will bring the users closer to ownership of their system. The whole process of developing a good systems model and preparing an honest cost-benefit report is clearly a very iterative one.

The increase in information realized in the creation of the Functional System Model allows the Version Plan and Cost-Benefit Statement to be refined. The Cost-Benefit Statement will no longer contain the major options it did earlier, but there will still be choices with respect to the extent of automation, and each such choice will have its own associated costs and savings.

Consolidating the Functional System Model

Consolidate the Functional System Model (process 3.3; see Fig. 67) is, again, essentially a packaging process to get the Information and Process models together—preferably with one data dictionary—in an acceptable format.

Sharing Models with Users

The Essential Business Model and the Functional System Model are not intended for dusty shelves in the data processing department only. These models should also be shared with the users, especially the Essential Business Model. For the same reason that architects deliver drawings to their clients, systems development professionals must deliver the architectural drawings to *their* clients: because they paid for them!

The Essential Business Model and the Functional System Model can certainly be used by line area managers to train new or promoted staff in the knowledge of the essential functions of their system and its interfaces with other systems or departments.

130   The Substance of the Model   Chap. 4

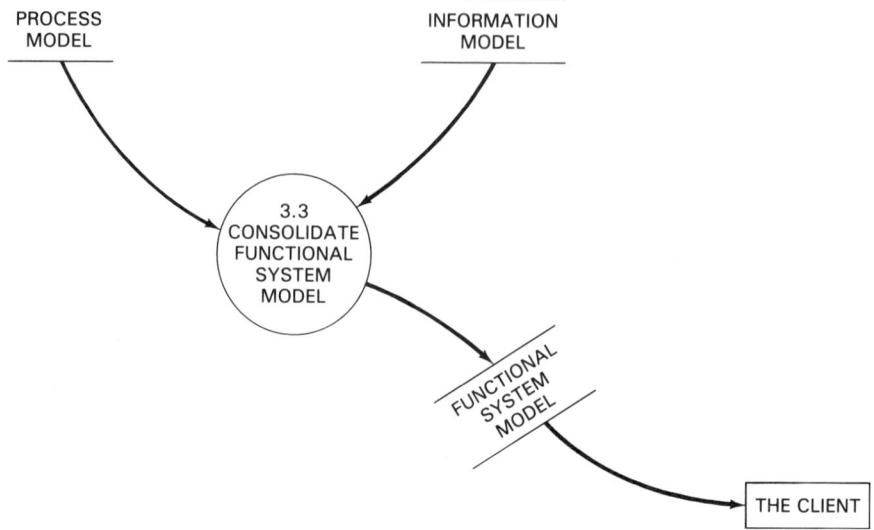

Figure 67: Consolidate the Functional System Model (3.3)

Acceptance Criteria

Finally, we should look at process 3.5 (*Record Functional Acceptance Criteria*), which does seem a bit unusual, sitting all by itself and with the Acceptance Criteria not being delivered anywhere. See Fig. 68.

Turn to the Dictionary entry for the definition of this deliverable. It's obviously a bit strange, since a single input produces a single output. Is there some unstated magic going on here? Is there a secret formula inside this process to transform the input, by itself, into something entirely different? No, of course not. The purpose of the process is to *record*, to gather

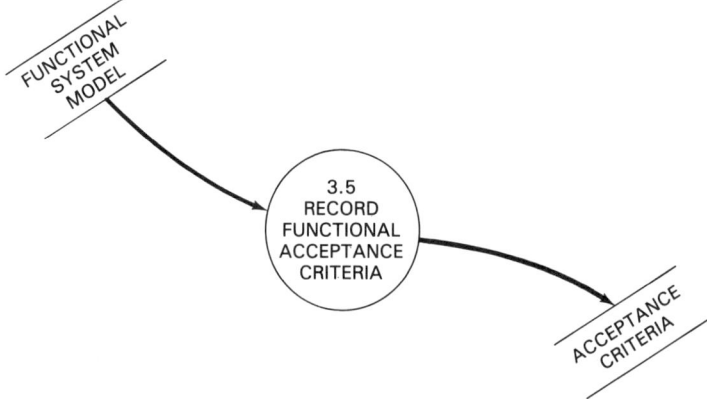

Figure 68: Record Functional Acceptance Criteria (3.5)

HUMAN RESOURCES SYSTEM
ACCEPTANCE CRITERIA LIST

1. There must be a facility for calculating compound interest on an employee's pension fund contribution (current year and previous years) on a per pay basis. Interest varies according to territory· see object definition PENSION INTEREST RATE as defined in the Data Dictionary. An employee can have contributions in more than one fund and in more than one currency.
2. Upon termination, company property held by the employee (including passwords and logon-IDs) must be displayed.
3. There must be a facility for the system to split-charge each earning for an employee to at least 6 work areas/branches. Each earning for an employee may be differently split-charged.
4. Before assigning an employee to a work area, there must be a facility to confirm the existence of that work area.
5. For START INCOME, the system must ensure that income name, income amount and/or overtime hours, start/stop dates, work area number, and split-charge percent are present and valid.

Figure 69: Human Resources System Acceptance Criteria List

information from one medium and put it onto another. Look at the Dictionary entry again. Bear in mind that this product will be revised with very physical criteria in Design. Although we are getting ahead of ourselves a little, this deliverable will be used to set up a testing strategy in Design, and then will be used by the Testing Team as a tool to know when the system has accomplished everything it was designed to do.

Although you may choose not to produce this document at this time, it will (when updated in Design) be much more useful than either the Process or Information model for the Testing Team to work with. Since it translates the requirements documented in the Functional System Model into a list of Acceptance Criteria, it becomes more of a left-brain tool, and therefore single-dimensional and less open to interpretation.

Also, when you are communicating with software vendors or systems consulting houses, the Acceptance Criteria can act very nicely as a specifications list, if walking through the complete functional specification with the many vendors is not practical. An Acceptance Criteria list might look something like the extracted example in Fig. 69.

Buying Packaged Software

At the end of functional specification—after the complete detailing of the client's function requirements—it is now possible to start considering the use of an existing software package or to start preparing Requests for Proposal (RFPs) to be sent to consulting houses and software vendors. Without the complete user specifications, the Development Team cannot intelligently discuss any modifications necessary to commercial software or accurately describe the impact of any given package on the client's business.

With a complete Functional System Model and a list of Acceptance Criteria, there are now objective standards against which commercial software can be measured.

An example of this is the preceding abbreviated Acceptance Criteria list, which resulted from a real development project. The criteria have been based on the Functional System Model but were stated textually so that software vendors could relate to them readily. The final list went on for some sixty-two pages. Note that audit and security acceptance criteria can be specified clearly and separately, although the control functions are integrated and distributed throughout the system.

The use of the Acceptance Criteria list and the Information Model (helps to show logical access paths to data classes) with software vendors was found to be so useful and informative that too much may have been discovered about the vendors' products. That is, the method identified not only the many strengths of the vendors' products but also made their weaknesses very apparent. Even making the decision to go with the best package proved to be difficult.

Also, because the Functional System Model clearly specified what the system was intended to do, the Development Team was able to deliver a meaningful Cost-Benefit Statement. There were several versions of these estimates delivered before the team and client settled on a palatable statement.

Figures 70 through 73 are examples of the kind of Cost-Benefit

Human Resources System: Cost-Benefit Overview

	Do Nothing	IOIs	Custom-Build System	Buy a Package
Operating Costs	5,187,087	4,907,931	4,506,676	4,414,555
Net Future Benefit	0	279,156	680,411	772,532
Present Value of Future Benefit	0	214,795	511,797	583,883
Development Cost	65,900	255,914	662,440	487,863
Present Value of Development Cost	65,900	226,623	570,166	431,414
Net Present Value of Investment	−65,900	−11,828	−58,369	152,469
Profitability Index	—	.95	.90	1.35
Return on Investment	—	6%	2%	30%
Delivery Date	—	end 1988	Feb 1988	Mar 1987

Figure 70: Cost-Benefit Statement Overview

Statement possible after a Functional System Model has been completed. They illustrate the summary page for all four standard options.

If, at this time in the systems development life cycle, a decision to purchase a commercial package is taken, the SuperSet model will have to be altered to reflect the different sorts of activities that will now be necessary.

The Bridge Between Analysis and Design

Before going on to the design process, we feel it's worthwhile to discuss further the importance the Functional System Model plays in the overall game plan of developing a good system.

For many years, at GUIDE and other forums, there has been a lengthy discussion of what is the bridge between analysis and design. It's possible to give several different answers to this question, primarily because of different meanings we assign to the words *analysis* and *design* in different contexts.

In a very practical sense there is no bridge, or fixed boundary, at all.

Throughout the systems development process, we gradually become aware of the probable physical realization of the system, and we continue working with that knowledge in mind. Some very physical details may be completely out of our control, such as the hardware to be used, the selection of the DS2 Multivariant Deeply Differential Operating System, or the choice of Bide'a'While Courier Service because it is owned by the client's nephew. We don't suppress the physical details that we know, and so we shouldn't. Design and analysis become *deeply intertwingled* as we begin to tie functions down to those physical hitching posts.

On the other hand, people haven't been talking about the "bridge" all this time for nothing.

There is a strong similarity between the processes *Create the Essential Business Model* and *Create the Functional Model*. In the first we are describing those activities that must take place for the business to exist, but with two major restrictions:

- we are ignoring technology (or assuming a technologically perfect environment);
- as well, all data is assumed to be ideal, available when needed, and completely clean.

In the Functional Model, not only do we go into more detail, but we drop the second assumption mentioned above. Here the processes to handle the problem cases, invalid data, and the technology at the system boundary must all be modeled, but still without any imposition of technology within the system boundary. In most cases the Functional Model will be an expansion of the Essential Model, with greater detail and with routines added to handle these issues of editing, boundary data conversion, and the like.

Human Resources System: Cost-Benefit Analysis

October 22nd 19xx

Summary IOIs	1986	1987	1988	1989	1990	Total 1986–1990
A. Operating Costs—current	820438	932362	1061509	1112660	1260118	5187087
B. Operating Costs—IOIs	820438	905321	996484	1026887	1158801	4907931
C. Net Future Benefit (A — B)	0	27041	65025	85773	101317	279156
D. Discount Rate (9.5%)	1	0.913	0.834	0.762	0.696	
E. Present Value of Future Benefit (C * D)	0	24688	54231	65359	70517	214795
F. Development Cost	65900	69740	75020	45254	0	255914
G. Present Value of Development Cost (F * D)	65900	63673	62567	34484	0	226623
H. Net Present Value of Investment (Sum E—Sum G)						−11828
I. Profitability Index (Sum E ÷ Sum G)						0.95
J. Return on Investment						6%

Figure 71: Summary: Immediate Opportunities for Improvement (IOIs)

Human Resources System: Cost-Benefit Analysis

October 22nd 19xx

Summary Custom-Build System	1986	1987	1988	1989	1990	Total 1986–1990
A. Operating Costs—current	820438	932362	1061509	1112660	1260118	5187087
B. Operating Costs—IOIs	820438	892362	972498	851025	970353	4506676
C. Net Future Benefit (A — B)	0	40000	89011	261635	289765	680411
D. Discount Rate (9.5%)	1	0.913	0.834	0.762	0.696	
E. Present Value of Future Benefit (C * D)	0	36520	74235	199366	201676	511797
F. Development Cost	65900	176931	319253	100356	0	662440
G. Present Value of Development Cost (F * D)	65900	161538	266257	76471	0	570166
H. Net Present Value of Investment (Sum E — Sum G)						−58369
I. Profitability Index (Sum E ÷ Sum G)						0.90
J. Return on Investment						2%

Figure 72: Summary: Custom-Build System

Human Resources System: Cost-Benefit Analysis

October 22nd 19xx

Summary Buy Software Package	1986	1987	1988	1989	1990	Total 1986–1990
A. Operating Costs—current	820438	932362	1061509	1112660	1260118	5187087
B. Operating Costs—IOIs	820438	901238	907959	836045	948875	4414555
C. Net Future Benefit (A — B)	0	31124	153550	276615	311243	772532
D. Discount Rate (9.5%)	1	0.913	0.834	0.762	0.696	
E. Present Value of Future Benefit (C * D)	0	28416	128061	210781	216625	583883
F. Development Cost	65900	286646	49266	42829	43222	487863
G. Present Value of Development Cost (F * D)	65900	261708	41088	32636	30083	431414
H. Net Present Value of Investment (Sum E — Sum G)						152469
I. Profitability Index (Sum E ÷ Sum G)						1.35
J. Return on Investment						30%

Figure 73: Summary: Buy Software Package

When we move on to design, however, there is a dramatic change. Now the *final* decisions must be made about what processors will be used, and how the activities illustrated in the earlier models will be allocated to these processors. In most cases this will mean that a different set of tools will be used; if data flow diagrams were the primary graphic tool earlier, now State Transition Diagrams or Structure Charts may predominate.

And this is the crux of the issue: moving from analysis to design isn't difficult at a conceptual level, but changing tools in midstream may be traumatic. Of course, it might be that you are designing for a network of parallel processors, and no great change is necessary. Whatever the tools used, however, it's important to keep clear the difference between the *processes* of analysis and design, which flow smoothly from one to the other, and the *tools* of analysis and design, which may be quite different and require an active strategy to move from one to the other. Such a transition strategy will help us "across the bridge"—and so we offer one as the first process of design.

4.2.4 Design the System (Fig. 74)

Most engineering professions have strong disciplines by which they develop their products. Software "engineering," however, has little discipline. Far too many DP systems are developed by a sulfurous mixture of mythology, witchcraft, and bad habits and hardly offer true creative satisfaction to a professional developer.[13]

One of the problems with the question of when should physical details begin to show in the system model is that we may have two different meanings for the word *physical* in the back of our mind:

1. very detailed user requirements are often referred to as low-level or physical requirements; this level of detail should be included in the Functional System Model;
2. characteristics of the system that describe how the system will interact with various processors, whether those processors are machines or people, are also called physical; these are design issues.

Design is the final placement of the Functional System Model in the context of physical processors. That's why the first issue we address is the allocation of functions to processors (Fig. 75).

The Functional Model is a *complete* specification of what the system must do, without any *final* decisions yet having been made as to how it will

---

[13] Meilir Page-Jones, *The Practical Guide to Structured Systems Design* (New York: Yourdon Press, 1980), p. 28. This is the first book to link the concept of structured design with the techniques of structured analysis. Designers working with process models will find Page-Jones' work invaluable.

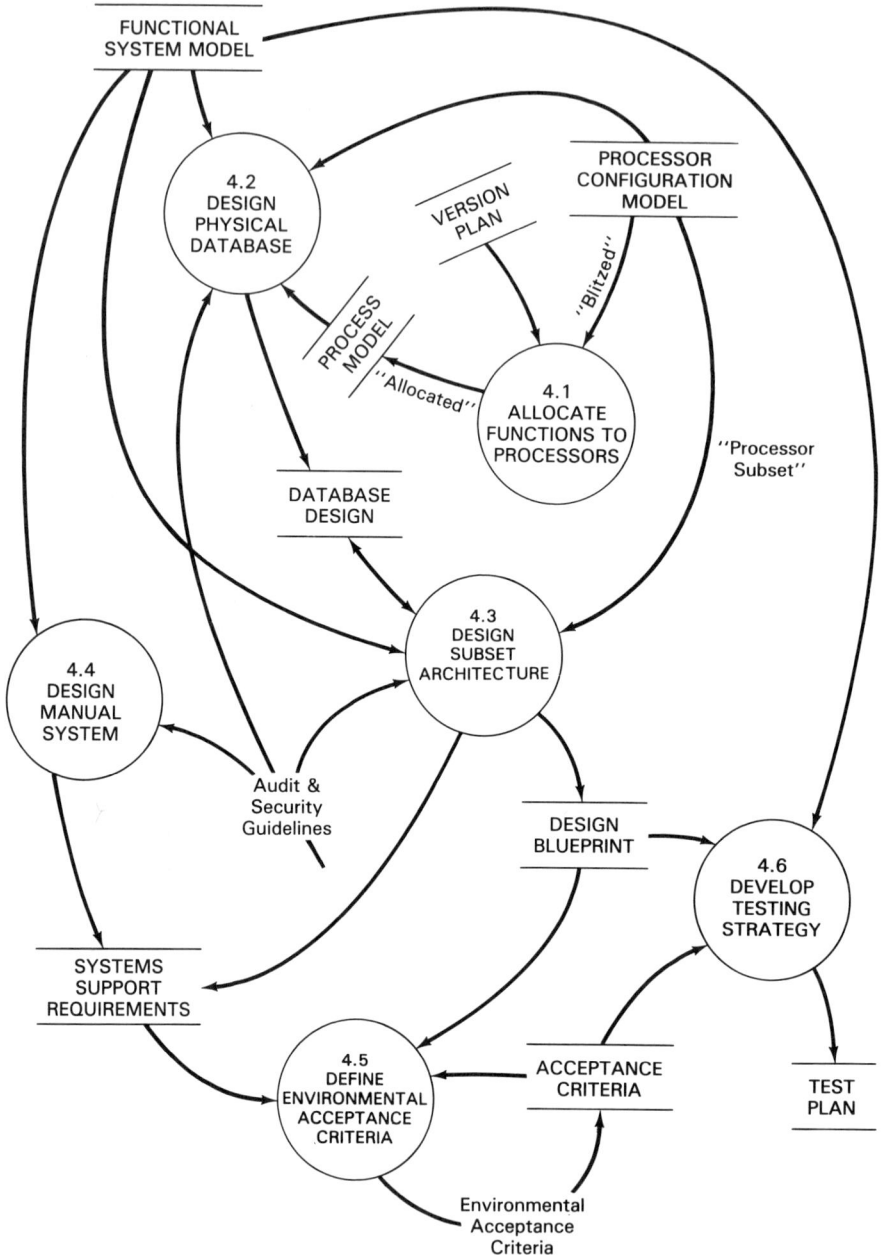

Figure 74: Design the System (4)

Sec. 4.2   The Detailed Diagrams   **139**

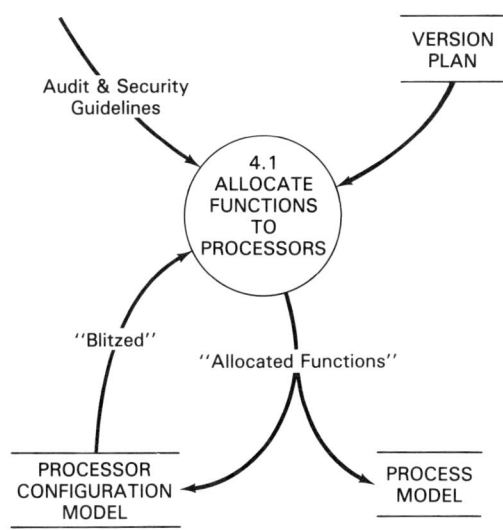

Figure 75:   Allocate Functions of Processors (4.1)

be done. It's also true that we've certainly thought about it a lot and played around with several possible alternatives with the user, so technology and costs wouldn't come as a complete surprise when we get this far into the project.

At the start of the design process a list of potential processors must be delivered. This list would include not only all possible hardware on which parts of the system could be implemented, but it should also specify those manual areas, or roles, that should carry out some of the tasks described in the system's Process Model. Some of this work has already been done by the users, in assigning the probable automation boundary (process 3.4.1, *Establish Automation Boundary*) and delivering a tentative Processor Configuration Model.

For the purpose of task allocation it's also useful to consider activities that will be happening in different time frames to be allocated to *logically different* processors, even if the same lump of machinery is involved. For example, a large mainframe could be considered to be the following machines for allocation purposes:

- the process control mainframe, for real-time activities;
- the interactive (clerical) mainframe, for commercial activities that must be processed on a routine and interactive basis;
- the year-end mainframe, for year-end processing;

Each activity of the Process Model must then be examined and allocated to the processor that is most suitable. Similarly, each data store must be assigned to a physical "container"; this may be a processor, or it may be a shared device between processors (such as a filing cabinet).

At the end of this allocation process, each processor subset of activities will have certain characteristics. These might include:

Processor
    human
    mainframe
    mini
    micro

Geography
    centralized processing
    distributed processing

Timing
    real-time (e.g., process control)
    highly interactive (online or frequent batch)
    periodic (month end, year end, etc.)

The criteria for this allocation process are the three Cs:

- Capacity
- Capability
- Cost

Each processor subset is now a jumble of activities taken from the Process Model. Actually, in most cases the content of the processor subset will be groups of logically related activities that have moved as clumps from parts of the Process Model. A series of leveled data flow diagrams must be produced for each processor. Because we are now dealing with a physical model, some groups of processors may be isolated from others within a processor.

Data flows must be added between the processors and any data stores outside of the selected processors. In many cases the only communication between processors will be through data stores.

As a last step in the allocation process, we must add to our model those processes that will handle the inter-processor communication activities, doing any packaging and unpackaging of data that is necessary because of the implementation that is chosen. A lower level data flow diagram for 4.1, *Allocate Functions to Processors*, might look something like Fig. 76.

The result of process 4.1 is an implementation model, or Processor Configuration Model.

Stephen McMenamin and John Palmer treat this entire subject extensively and extremely well. Processor configuration for any reasonably

Sec. 4.2   The Detailed Diagrams   141

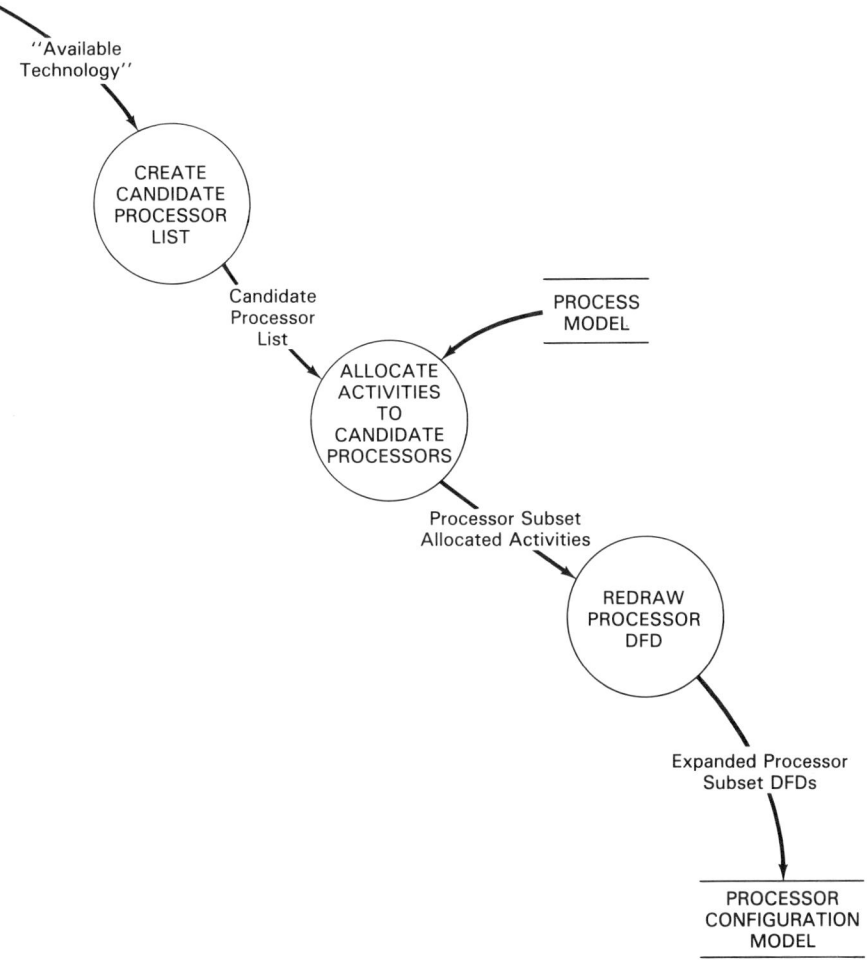

Figure 76: Allocate Activities to Processors

complex system is in itself a complex subject, and the how of it deserves the detailed treatment McMenamin and Palmer give it.[14]

Each processor subset can now be viewed for the purpose of further design and construction as an independent subsystem, which facilitates the allocation of resources. The further design of each subsystem is modeled in Fig. 77, *Design Subset Architecture* (process 4.3).

This is the meat and potatoes of design and requires a variety of design skills, as may be indicated by this very busy process. Let's decompose it some more, as in Fig. 78, so we can get a handle on it.

[14] McMenamin and Palmer, *Essential Systems Analysis*.

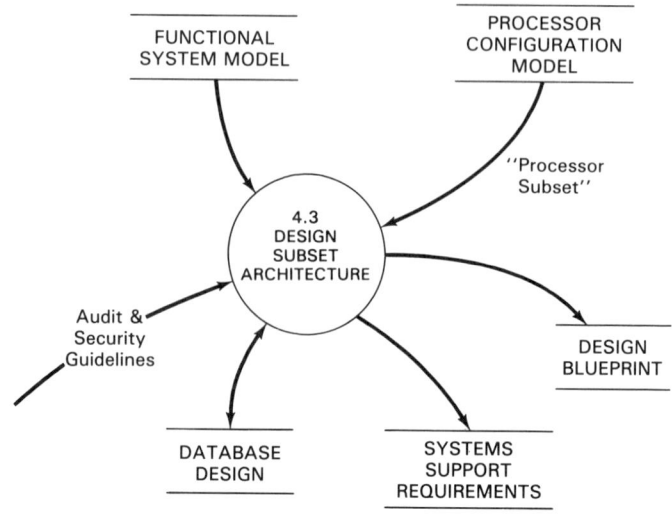

Figure 77: Design Subset Architcture (4.3)—Level 1

Since your internal Audit & Security Guidelines may influence every process at this level, we have not shown it on the level 2 diagram, since it would only serve to clutter the picture and not communicate anything new. Again, consideration and application of things important to the audit and security functions are always going on in the background.

If the client's system has an online function, then a dialogue will have to be designed (process 4.3.2). The tools used in Dialogue Design and the creation of scenarios will strongly influence the rest of the design and construction processes, and the life cycle model will have to be modified accordingly.

It's important that the dialogues be specified early in the design process. (In fact, the dialogue is analogous to a report or input document and could even be considered much earlier in the development life cycle, as long as it is clear that there are no physical alternatives to an online implementation.) Certainly the client will have strong opinions on screen composition and the flow of the dialogue, and these must be resolved before consideration of other online dialogue issues.

Dialogue Design can be facilitated through the use of prototyping tools that allow the user hands-on experience with the dialogue as it is being created. If such tools are not available, "static" screens can be mocked up on a terminal or microcomputer, which will show the user what the dialogue will look like but without permitting any interaction. As a last resort, paper screen layouts can be produced. But these are so distant from what the user will eventually have to work with that any commitment a user makes to screens presented this way can't be fully depended on once the user has seen

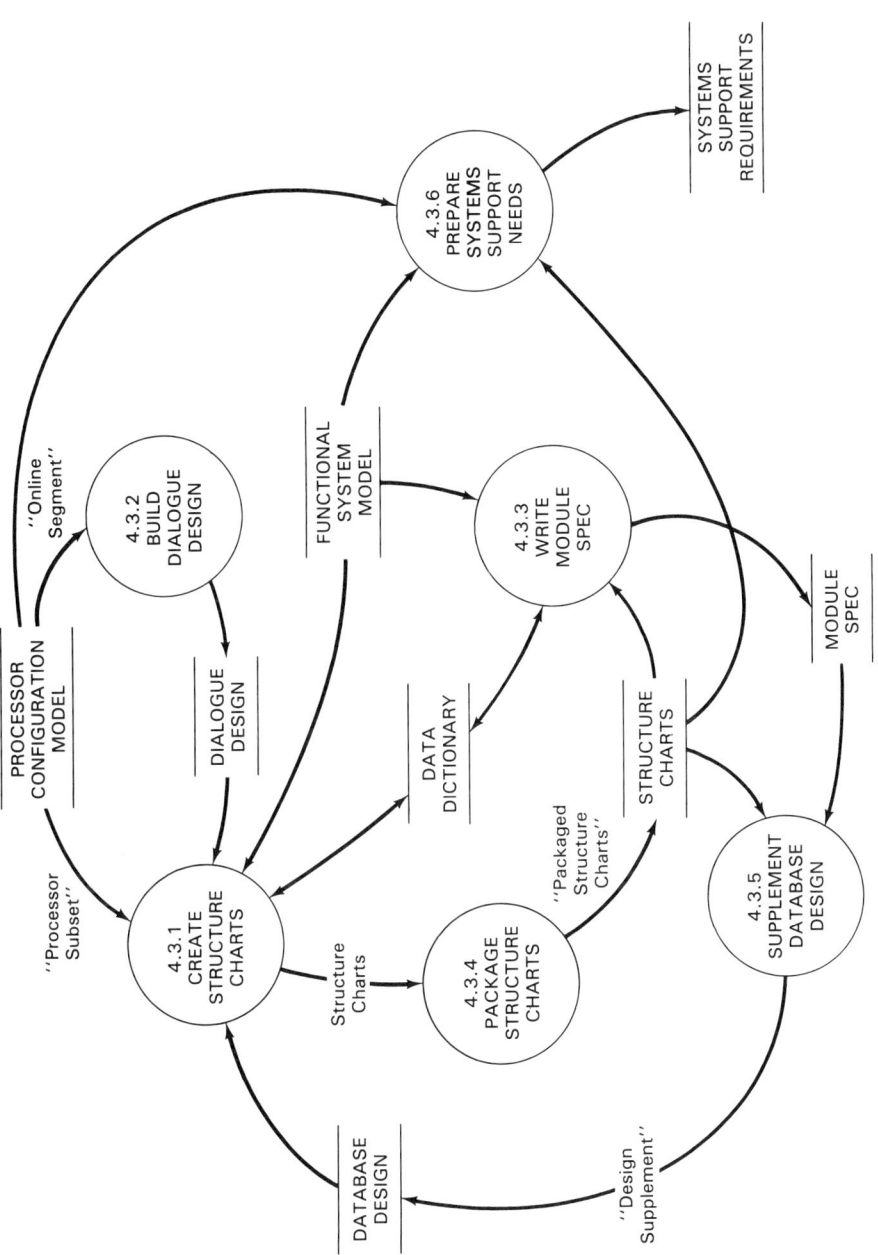

Figure 78: Design Subset Architecture (4.3)—Level 2

the real system. In doing any dialogue prototyping, it's also prudent to build delays into the pseudo-dialogue that will reflect your best estimate of the actual response time the user will encounter—it's always safe to have the real system seem faster than the prototype when implemented, but politically unhealthy for it to seem much slower.

Once each panel in the dialogue has been finalized, we must face the question of how to implement the dialogue, and what it means to "design" the implementation.

In a generic sense, there will be three categories of process routines necessary:

1. a routine to control or manage the dialogue flow;
2. a "predisplay" routine for each panel, which will ensure that all data needed to put up each screen are available and in a suitable form;
3. a "postdisplay" routine for each panel, which will edit and cross-edit the data entered on a screen and make certain (Boolean) variables available to the dialogue manager, which can then make any necessary decisions on the flow of the dialogue.

Depending on the tools being used, not all of these types of routines may be necessary; some dialogue prototyping tools act as a dialogue manager in a production environment; other systems use table-driven dialogues (with the actual dialogue manager being the software that interprets the table). In many cases the pre- and postdisplay routines will be null sets, if no special processing is necessary.

Certainly in lower level languages, we have found it most useful to think of the dialogue manager as a giant case construct, switching on screen name. The dialogue manager should know nothing about the application, and the application routines should know nothing about the dialogue flow. This approach to dialogue design quickly breaks the online subsystem into a series of processor subsets for the sake of detailed design. This approach to decomposition produces very quickly a group of highly cohesive routines. By contrast, viewing the panels as I/O operations comparable to reads and writes doesn't lead immediately to as nearly a fruitful decomposition.

Each processor subset, whether online or batch, will have a Structure Chart created for it (process 4.3.1). Then (in process 4.3.4) the components of each Structure Chart will be allocated to specific implementation packages such as JCL procedures, programs, code procedures, functions, sections, paragraphs, and so on. The processor subsets of an online system may not require much further decomposition, except perhaps into language-dependent units, such as procedures, source files, or paragraphs.

Module specifications (done in process 4.3.3) must be written before or concurrently with the Structure Chart packaging process. Meilir Page-Jones's book is possibly the most readable outline of the "structured" approach to detailed design.[15] However, the "data-structured" approaches of Michael Jackson,[16] Jean-Dominique Warnier,[17] Ken Orr,[18] and others can be very powerful in many cases.[19] Also, *Structured Design* by Ed Yourdon and Larry Constantine is a classic in software engineering and should be read by everyone.[20]

Because Structure Chart packaging by processor subset may dictate additional or new data needs to facilitate communication among processors, the Database Design may have to be supplemented (process 4.3.5, *Supplement Database Design*). For example, if two physical programs were used to implement one processor subset, a new file may be necessary for the data flow between the programs, as well as new procedures to read and write the file. New physical data requirements would include the file itself, plus all the accompanying procedural data (such as "end of file" flags) in each program.

Also, after the procedure of designing the subset architecture has been completed and the "physical" things needed to build and operate the automated side of the new system have been identified, the Development Team can gather together much of the Systems Support Requirements (in process 4.3.6). At this time, this deliverable specifies the hardware and production environment support software needed by the system.

Meanwhile, back in Fig. 74 (*Design the System*), a variety of other processes will have been carried out, or at least started. When creating your own superset approach each of these processes should be modeled to at least one lower level of detail. We haven't shown lower levels here because at this level of physical detail everyone's model will be different. All of these processes are very project- and environment-dependent (such as processes 4.2, *Design Physical Database*, and 4.4, *Design Manual System*). When the manual system has been entirely designed, the Systems Support Requirements can be completed to reflect the physical needs of the new system outside the automation boundary, including the forms needed and the personnel needed to support the system.

During the process of designing the new database (process 4.2) it's necessary to notify others of required changes to existing database files. Unless a formal data administration group exists in your company, other

---

[15] Page-Jones, *Practical Guide to Structured Systems Design*.

[16] Not the musical superstar but M. A. Jackson, the brilliant author of *Principles of Program Design* (New York: Academic Press, 1975).

[17] Jean-Dominique Warnier, *Logical Construction of Programs*, 3rd ed. (New York: Van Nostrand Reinhold, 1974).

[18] Orr, *Structured Systems Development*.

[19] The best introduction to the concepts of "data-structured" design is perhaps Kirk Hansen's *Data Structured Program Design* (Topeka, Kan.: Ken Orr & Associates, Inc., 1984).

[20] Edward Yourdon and Larry L. Constantine, *Structured Design* (New York: Yourdon Press, 1978).

departments, development teams, or systems will need to know the following and often more:

- *new data definitions* (i.e., role, meaning, purpose, function, source);
- *data content* (i.e., value restrictions);
- *data structure* (i.e., alpha/numeric, precision, units);
- *data dependencies* (i.e., dependency on other data elements for existence);
- *authority to change*
- *when the change takes effect.*

The Acceptance Criteria can now be completed because the "physical" components of the system have finally been clearly defined. The Acceptance Criteria, done at functional specification time (process 3), reflect the functions that the client insisted on receiving before she would accept delivery of the system. The Environmental Acceptance Criteria details the requirements (in process 4.5) that relate to the processor environment such as response time, throughput rates, security features, recovery procedures, and so forth.

Testing Strategies

This leaves process 4.6, *Develop the Testing Strategy.*

Because all the system's functions have been defined, all the processors have been allocated, and therefore the Acceptance Criteria have been completed, the testing strategy can now be properly developed. A caveat, however: The Test Plan should be developed by members of the client's staff external to the project rather than the Development Team, if at all possible. The Development Team, firmly convinced of the grace, elegance, and correctness of their work, will have great difficulty in creating tests to catch the flaws that they believe do not exist!

Finally, because everything has been nicely and unambiguously specified, the Development Team (or external consultants) can now fix the cost of completing the delivery of this version of the system to the client.

This is the first time we can do this without holding our fingers crossed behind our back! Although the cost estimates done after the *functional specifications* were probably pretty good, they were still likely only within 25 percent of the real numbers. Many other systems professionals may disagree with this. Some have clearly defined metrics for predicting project development costs in the very early stages of process modeling. Few, however, consider the specific allocation of processors to processes and what this does to the amount of work needed to complete the system. Processor allocation can be a fairly significant contributor to the completion cost, and is most certainly difficult to assess when the functional specification is delivered; that is, while everyone is still speculating what processors will be employed, and how.

Sec. 4.2 The Detailed Diagrams   **147**

Although many things happen in design, the major deliverable of this process is the Design Blueprint, which consists of *packaged* Structure Charts; a complete Data Dictionary; the Module Specifications; and, if necessary, the Dialogue Design, which could include State Transition Diagrams.

4.2.5 Build the System (Fig. 79)

Never again shall a maintenance programmer squander the gold of his youth on a system that is inherently unmaintainable.

Meilir Page-Jones[21]

In the dark ages of systems development (that is, prior to 1975) the writing of code often started quite early in the development process. Managers who had been shuffled into a DP department because of successes (or failures) elsewhere viewed the creation of compilable code as the primary function of their department. The legacy of those early projects are our multiyear maintenance backlog (which has nothing to do with lack of good programmers) and the lack of confidence many users have in their data processing sections.

With the use of the structured techniques, the process of writing the code becomes a blip of activity late in the development schedule. One of the authors (Frantzen) consulted to a project in which coding was about 11 percent of the entire development effort, while testing was another 11 percent. (The development team was not aware of the measurement.) The project team used the superset concept and devoted 7 percent of their time to developing a new Essential Business Model (one did not exist previously), 28 percent creating a Functional System Model, 35 percent on the Design Blueprint, and 8 percent on training and implementation. Conversion in this case was trivial, consisting of less than 1 percent.

Perhaps, in a few short years—with the more powerful languages and automated tools of the coming fifth generation—coding will become a thing of the past and design documents will become "compilable."[22]

There are three major spheres of activity in the construction process: *build* (process 5.2) *test* (processes 5.1 and 5.3) and *document* the system (processes 5.4 through 5.7).

The process of testing (Fig. 80) is very project-dependent, but should certainly be modeled to a lower level.

As mentioned in the previous section, testing should be carried out by a team independent of the Development Team whenever possible. To this end we have defined a Testing Team and its preferred attributes in the

[21] Page-Jones, *Practical Guide to Structured Systems Design*.

[22] The coming future generations of computer languages and artificial intelligence is discussed extensively—past, present, and future—in *The Fifth Generation*, by Edward A. Feigenbaum and Pamela McCorduck (Menlo Park, Calif.: Addison-Wesley, 1983).

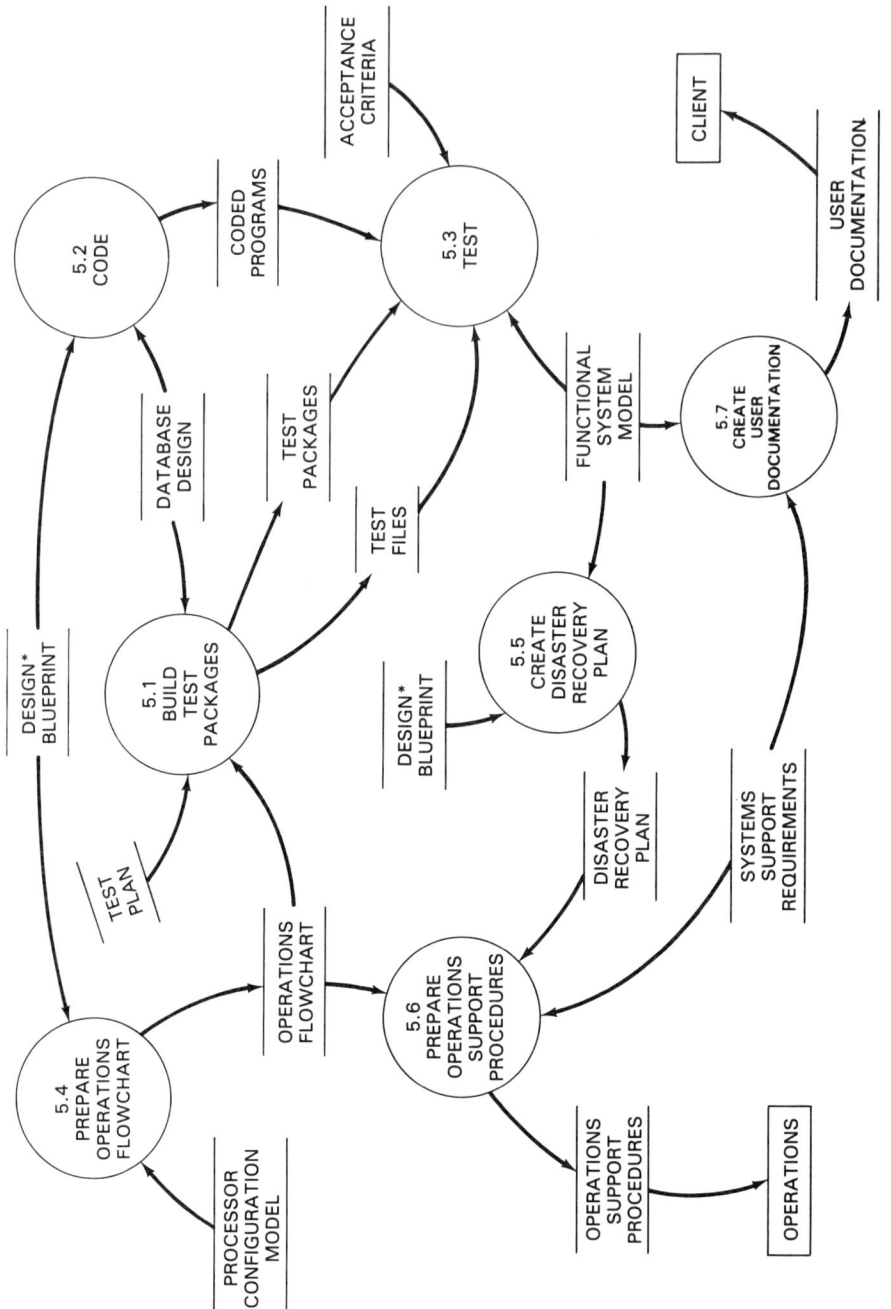

Figure 79: Build the System (5)

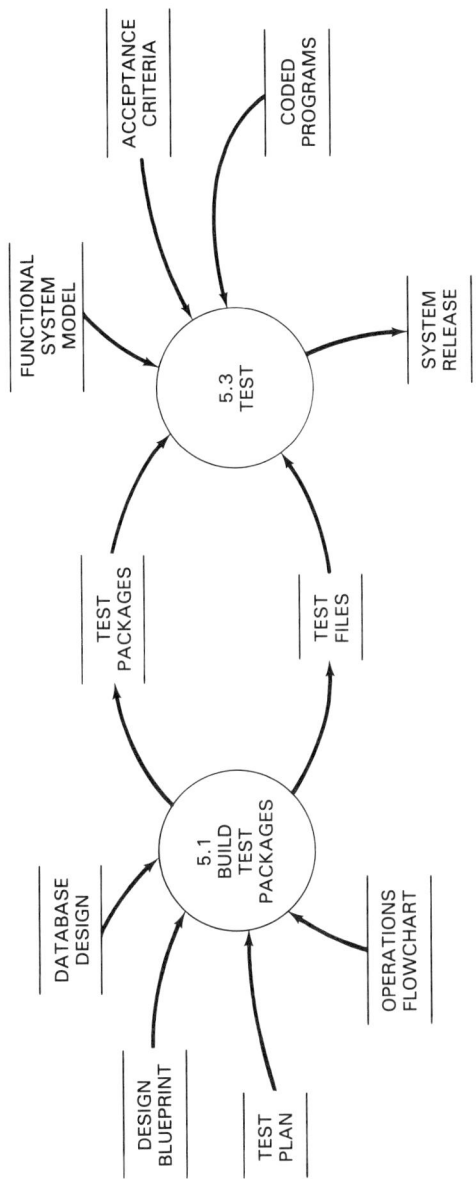

Figure 80: Testing the System

Dictionary. Also, several valuable works on testing are available for further investigation of the field. A lot of books have been written on testing during the past decade. Most are out of date. Two, however, still stand out from the others: Glenford Myers' *The Art of Software Testing* and Robert Glass's *Software Reliability Guidebook*.[23]

The Operations Flowchart (or *systems flowchart*) is produced at the beginning of construction (in process 5.4, *Prepare Operations Flowchart*), never earlier, because it's only at the very end of (a cohesive segment of) design that we allocate tasks to physical packages such as programs.

Before documentation can be produced for the Operations group, a final deliverable must be prepared. The Disaster Recovery Plan, produced in process 5.5, describes for the Operations staff and users what to do when disaster strikes. What this really means is: What do we do to restart operations after the loss of the entire data center facility?

Consideration must also be given to the impact of a disaster on the manual side of the system. Are any data vulnerable by being stored on paper in one location only? Are there legal documents or contracts that require special treatment? How would the business process be affected by a loss of communication facilities?

If this deliverable does not exist, it is possible that the entire business served by the system under development will fail within a matter of days or weeks, perhaps causing the total failure of the company.[24]

The only "documentation" produced, in the traditional sense, is that which is needed to have Operations run the system in production (process 5.6, *Prepare Operations Support Procedures*) and to show the users how to use their new system.

Create User Documentation (process 5.7; Fig. 81) is often treated by a lot of systems developers as rather unimportant. Sometimes, the approach they take is very much "Well, if we have time we'll see what we can do. We have real work to do, you know, like maintaining programs!" To paraphrase Glenford Myers: "If you don't have time to do it right, you sure don't have time to do it wrong!" Nontechnical User Documentation is an essential part of the system, and *must* be delivered.

Although it would be easy to believe that the User Documentation will take months to deliver, and that after the User Documentation there will still be a tremendous volume of systems documentation to produce—a vision of horror only partly offset by the fact that these tasks will ensure perpetual job security for an expanded word processing department whose tenure other-

---

[23] Glenford J. Myers, *The Art of Software Testing* (New York: John Wiley, 1979); and Robert L. Glass; *Software Reliability Guidebook* (Englewood Cliffs, N. J.: Prentice-Hall, 1979).

[24] Louie A. Molnar has written an exceptionally good article, entitled "Recovery Control Standards" (EDP Auditor, Fall 1983, pp. 11–18), which can be used as a basis for developing backup, restart, recovery, and control/security standards in just about any environment.

Sec. 4.2   The Detailed Diagrams   151

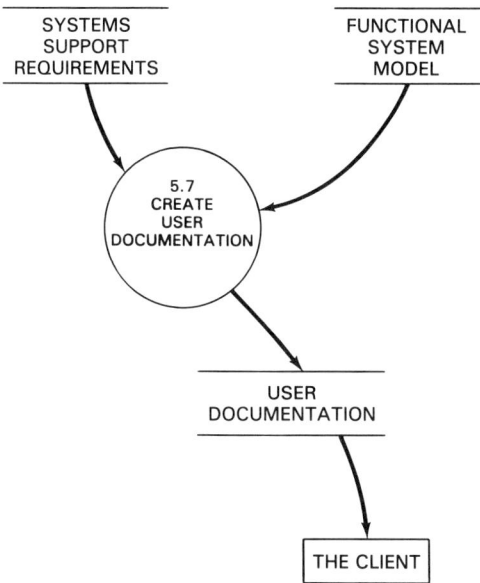

Figure 81: Create User Documentation (5.7)

wise seemed threatened by the encroachment of micro-based word processing packages—take heart; it just isn't so!

This deliverable should be small. The user does, after all, get a copy of the Functional System Model, which in itself is a clear graphic and textual description of all the functions that happen in the user's area and that specifies all the rules as well. The User Documentation package, on the other hand, describes all the user's procedures that are part of the system (inside and outside the automation boundaries), but which are not usually defined in sufficient detail in the Functional System Model. These are detailed descriptions of clerical activities and procedures, and what to do when. There is a full definition in the Dictionary.

One way around this nagging problem is to get the users to create their own documentation. They have, after all, been working closely with the team throughout the development process, so they should have no problem producing this deliverable.

Remember the second cardinal rule of the game plan approach:

* BAN EXTRANEOUS DOCUMENTATION *

Specifically, ban DP documentation. The only documentation needed for the data processing department (which may not even have been involved in developing the system!) is the documentation created in the development process itself. That is, the Essential Business Model, the Functional System

Model, the Design Blueprint, and a few other loose ends. (What's kept is defined in the Dictionary under Stuff Kept and Maintained.)

No resource should ever be allocated to task that don't move the project closer to completion; and after-the-fact documentation is one of the most demoralizing, boring activities imaginable. Period. The only exception to this rule is when the systems developer is contractually obligated to provide certain kinds of documentation. Government contracts are the most obvious example, and the Mil Spec documentation required by the U.S. Defense Department is probably the most onerous example. (The U.S. military establishment is not alone in this bureaucratic boondoggle; their military counterparts in Canada, the United Kingdom, and elsewhere take a similar approach.)

4.2.6 The Conversion Process

The conversion process is a smaller version of the entire life cycle model (see Fig. 82).

From the Version Plan, it can be established what changes to data are necessary for a particular implementation. This change requirement is analogous to the Request for Service that starts the entire project, and the **Conversion Scope** (delivered from process 6.1) is just a Preliminary Agreement for conversion. Similar analogous procedures of modeling, design, and construction should lead to the "implementation" of the conversion system, which collects or converts the existing data from their current format to that needed by the functions being implemented in this version of the target system.

4.2.7 Develop Training Materials

In the *Develop Training Materials* process all the materials necessary to train the client and user community and the Operations team are developed (see Fig. 83).

Before this is actually done, however, there is a whole procedure we might call "Design the Education Program" that has to be done. Again, depending on the needs of your company, you may want to see this process as part of design (process 4, *Design the System*) or training (process 7, *Develop Training Materials*).

Before packaging the Operations or Client Training Package we have to discover the *actual* need (distinct from *perceived* need), the preferred format, costs involved, schedules that have to be met, and so forth.

All of the following, and possibly more, have to be considered before training packages are put together:

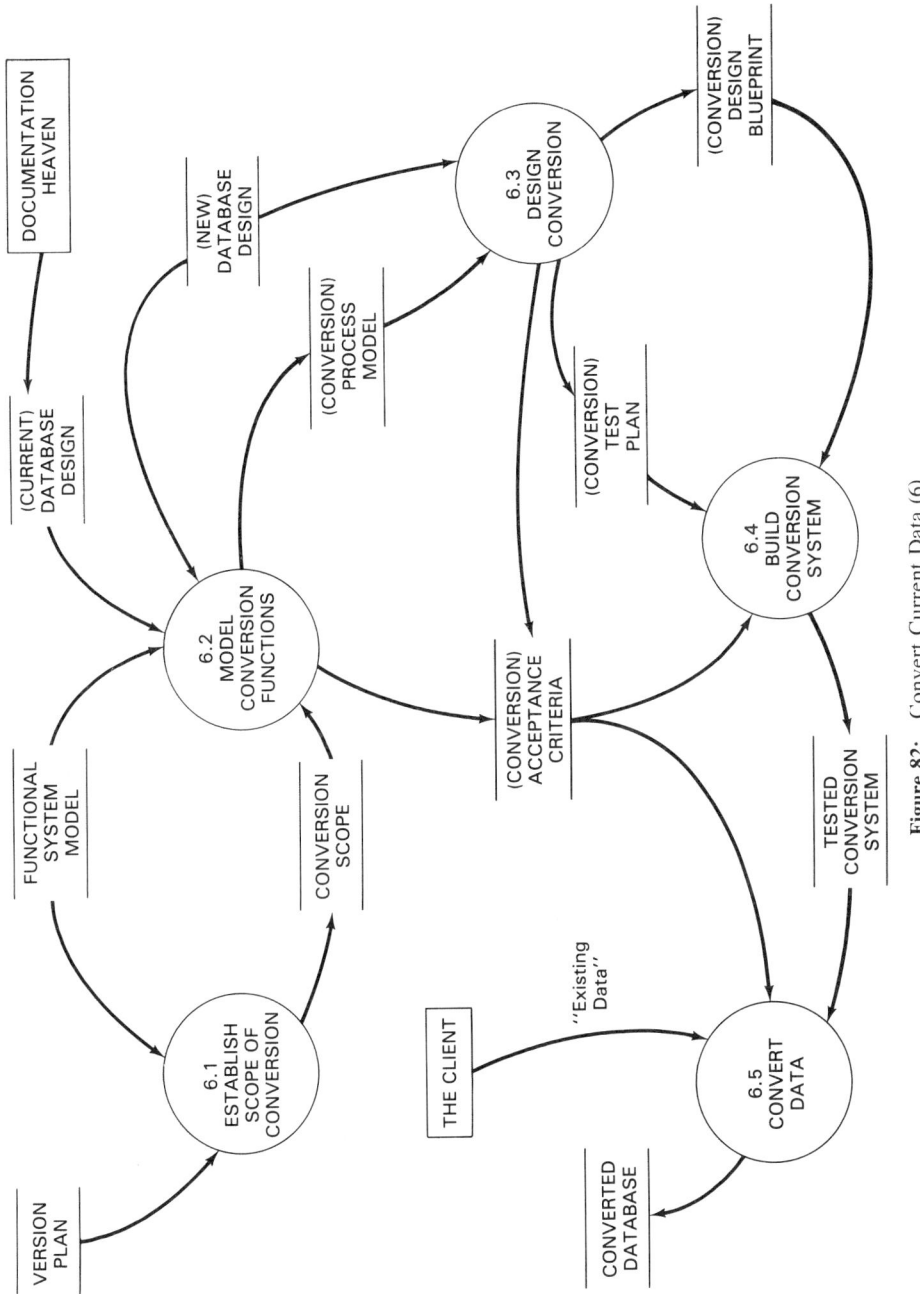

Figure 82: Convert Current Data (6)

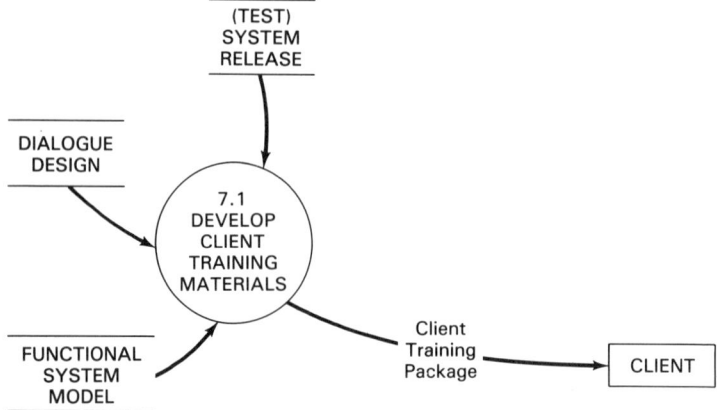

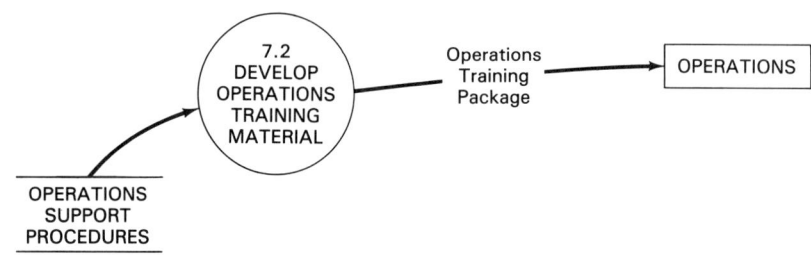

Figure 83: Develop Training Materials (7)

- How big is the audience?
- What is their experience level?
- What do they need to know about the system in order to do their jobs?
- How can the system's documentation be used to help them understand the system?
- What format is best suited to the audience (e.g., classroom, self-study, on the job with a mentor during pilot mode)?
- What external resources will be needed (e.g., audiovisual, consultants)?
- What is the estimated time and cost for development of the training packages?
- What is the estimated time and cost of conducting the training program?
- What is the delivery plan and schedule for:
  - equipment needed?
  - classrooms and other facilities?
  - preparation of student materials?
  - "train the trainer" sessions?

Sec. 4.2   The Detailed Diagrams   **155**

Regardless of the process of creating user training, the primary deliverable for users of the system from this process should be a series of short, concise "cheat sheets" that explain the interface dialogue with machines.

The client's clerical staff must, by necessity, be well trained and be able to understand and do their work. But since most people don't carry a training package with them forever on the job, the Client Training Package should simply be a reminder of *most common things*, or a series of *cheat sheets*. Everyone should have his or her personal copy. Things that sit on a shelf or in the library don't get used very often. This deliverable is not intended to be very big or complicated.

One area that can be big and complicated, however, is the use of online help screens. Once again, these should be developed with the user and the help of a communications specialist. Also, help screens should be integral to the system and created as screen dialogues are developed. For the online segment of the system, this may be all the training documentation you will need.

In our example model, the training material is delivered to the client and operations group, and the actual responsibility for delivering the training is external to the project. However, an alternative model could be prepared that showed the Development Team involved in the delivery of training.

4.2.8 Implementation

Implementation (Fig. 84) begins with the unavoidable bureaucracy of delivering the System Release into the hands of Operations.

All material that has been produced during the development cycle that will be of value to future Development Teams must be assembled and delivered to Documentation Heaven. (See the Dictionary for Documentation Heaven and Stuff Kept and Maintained.) This includes the program code, although individual security requirements may demand that code be stored in a more conventional location as well.

Most projects will have a period after implementation during which the behavior of the new system is monitored closely and compared against the Acceptance Criteria (8.2, *Monitor Shakedown Results*). If the implementation involves parallel runs, output comparisons between the old and the new systems need to be done. Any problems or discrepancies between observed behavior and the Acceptance Criteria will result in Required Change requests being generated for consideration in the "maintenance" cycle.

With the end of implementation, the development process is over, and the Development Team can fold up their bubble templates and silently steal away.

But not until a Postproject Review has been held!

Postproject Reviews (better known as PPRs) are vital, because only through examining the experience of the team can people really learn from

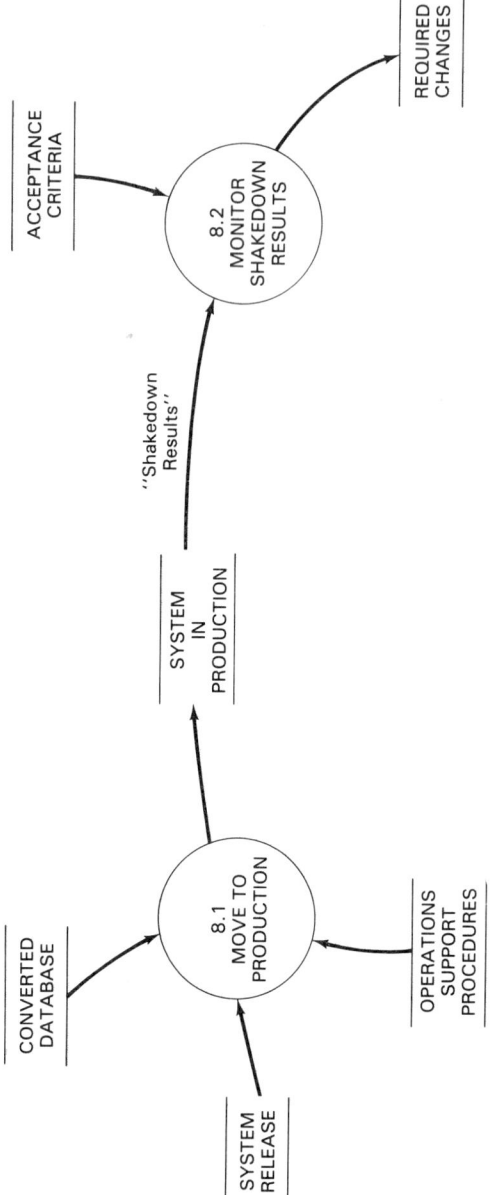

Figure 84: Implement the System (8)

their mistakes or reinforce those aspects of the project that went very well. Companies that don't do PPRs are doomed to develop systems basically the same way every time, whether it works or not! Among other things it's the Postproject Review that feeds "stuff going on in the background." It's also an opportunity to provide feedback to the team so they can improve their estimating skills. If you don't do PPRs, your people won't grow, and they certainly won't have any base on which to improve their estimating skills.

Also, the client should get a Completion Report after implementation. We haven't specified this in our example of the SuperSet model, primarily because it's a management activity rather than something that helps to deliver the target system. But the client is probably interested, and the report should discuss some of the following:

- the degree of success attained by the new system, in terms of meeting the Acceptance Criteria;
- the objectives in the Acceptance Criteria *not* attained;
- an action plan to upgrade the system to meet the Acceptance Criteria not attained;
- a refined Cost-Benefit Statement based on actual development, conversion, and implementation costs;
- a comparison of the actual (completed) cost to develop the system and the estimates made with the delivery of the Functional System Model—and a promise to estimate better next time!

Figure 85 illustrates what the Postproject Review process might look like. The Estimating Database is a collection of all estimating metrics from previous projects in the organization. If, however, there is never any "estimating feedback" provided to the Development Team and the feedback is never formally collected, then the Estimating Database will never exist, except in someone's mind. Without a formal collection of previous metrics, the estimating exercise will never become rigorous, but rather will be a matter of thrusting a wet finger into the air and guessing wildly at what our experience, or lack of, will bring us. Meeting schedules, deadlines, and estimates could then very well be a matter of determination and beneficial wishful thinking, rather than some measure of science.

4.2.9 Maintain the System

The project may be over but the life of the system is just beginning. Now the responsibility for the system passes from the Development Team to the *Quality Management Team*.

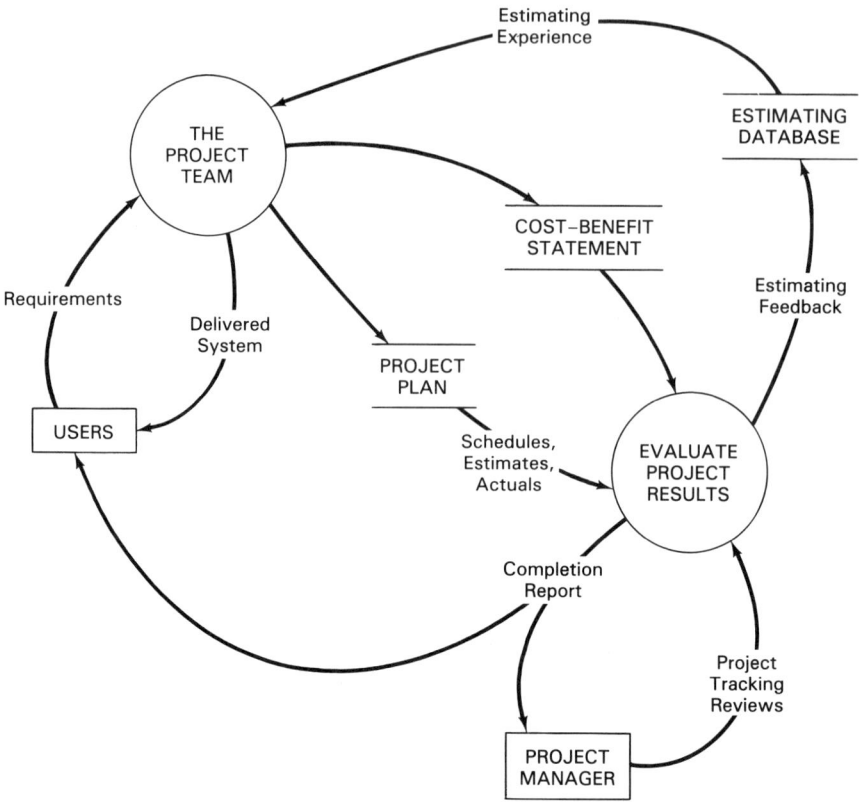

Figure 85: Evaluate Project Results

In one study of the development life cycle[25] it was found that 67 percent of the lifetime costs of a system occurred during the maintenance part of the life cycle. This also suggests that there is a tendency by some people in the systems building business to not treat the maintenance issue very seriously. Although there is a great deal of interest in systems development life cycles, many managers insist these methodologies are only for new development and only for big projects! Some of them also insist that the systems life cycle ends with implementation—just when the system has been born! Yet, in most mature companies today, it takes closer to 75 percent of the total DP resources just to keep old and somewhat rickety systems up and running! This sounds like a lot of people are spending a whole lot of money trying to find a methodology they can use perhaps 25 percent of the time, while completely ignoring the maintenance segment.

[25] M. V. Zelkowitz, "Perspectives on Software Engineering," (*ACM Computing Surveys*, June 1978, p. 202).

Sec. 4.2   The Detailed Diagrams

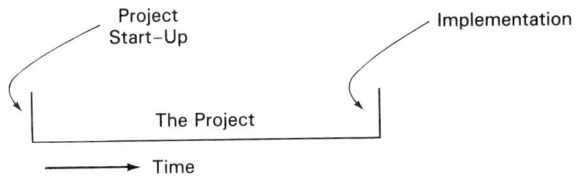

Figure 86:   The Typical Systems Life Cycle

A systems development life cycle does not mean what is depicted in Fig. 86.

Robert Block said in *Politics of Projects*,[26] "Organizations do not need projects: they need systems." It is clearly the responsibility of someone to professionally feed, care for, and continue to develop the system we have just given birth to. Another way of looking at the systems life cycle, then, would be what is diagrammed in Fig. 87.

To address the complete systems life cycle, including "enhancement," we have introduced the concept of a Quality Management Team, perhaps in a sense different from the norm. This is the team of people associated with the implementation and quality management of a client's system. This team also feeds and cares for the system after implementation. They were previously known as "maintenance programmers," but this is far too limiting a role.

In some companies it's possible that the Development Team and the Quality Management Team consist of the same people, or at least have some crossover.

The roles of the Quality Management Team could be to:

- prevent degradation of a system's "goodness";
- maintain and give a conscience to quality standards;
- measure quality;
- participate in Postproject Reviews of results and methods to produce results;
- evaluate the techniques in use;
- improve quality accounting methods;
- establish and publicize quality standards, but only as guidelines;
- specialize in current technology and how it can be used to solve business problems;
- develop and maintain expertise in
  - documentation
  - database
  - telecommunications

[26] Robert Block, *The Politics of Projects* (New York: Yourdon Press, 1983).

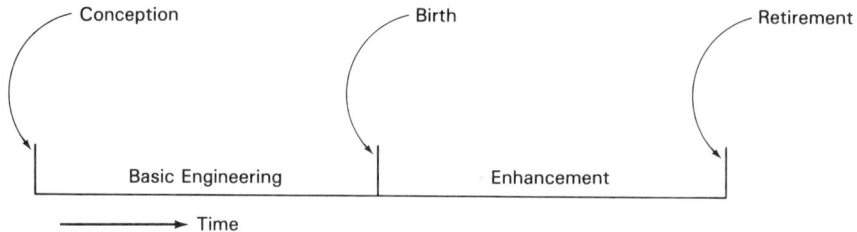

Figure 87: The Complete Systems Life Cycle

- testing
- programming
- be responsible for the coding and implementation of systems;
- control the System in Production;
- be the client liaison re: production;
- be advisers to Development Teams on technical matters;
- have an active role in quality assurance on development projects.

The Quality Management Team should never dictate methods, measure people, or standardize any untried techniques.

The Maintenance Conundrum

The traditional view of "maintenance," comprising simple code fixing and perhaps some documentation updating (but not likely, since it's hopelessly out of date anyway!), looks like the diagram in Fig. 88.

By expanding this to a more organized and methodical approach, we

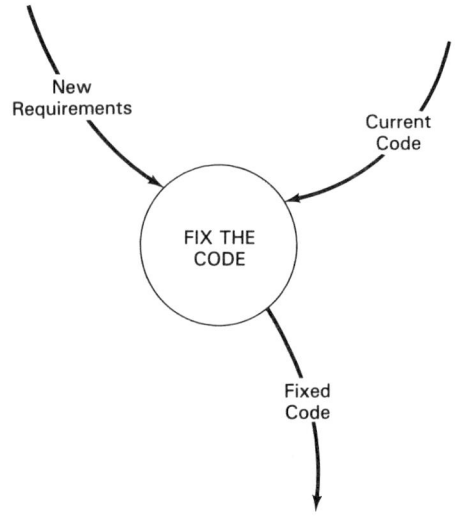

Figure 88: Traditional View of Systems Maintenance

could very well develop systems that are truly enhanced, and not simply grafted on to. This could lead to a maintenance activity that is truly productive and can be seen to be so.

Although maintenance can be clearly seen as part of the system's life cycle, no Development Team will ever model the maintenance process as part of their project life cycle; they will assume the system they are about to build is perfect and never needs changing! (They're realists—they know this isn't true—they just don't want to face the maintenance issue, since it's well known that whoever will maintain their wonderful system certainly won't maintain the documentation!) On the other hand, the Quality Management Team isn't interested in the original development process, though they will use output from that process (Stuff Kept and Maintained). The Quality Management Team will create new processes from scratch, viewing any Required Changes or Requests for Service as the original stimulus into process 1, *Set Up Preliminary Agreement*. So, even though no breakdown of process 9 is needed, we will offer two gratuitous models to show the role of maintenance in the entire life cycle.

The major principle behind the superset approach is the idea that from a superset of all possible activities, only those activities that are necessary to a project's completion are selected, built into the project's process model, and carried out. In maintenance (or systems enhancement) this is much more important than in development because it's possible to leave out so much more.

Earlier we pointed out that change requests can come from a variety of sources and be of varing degrees of complexity. But, no matter where the changes come from, what causes them, or the degree of complexity they bring with them, there are essentially only two kinds of change requirements:

1. changes to *essence* of the existing system; or
2. changes to the *implementation* of the existing system.

Figure 89 illustrates that every change request should be measured against each systems model, from the Essential Business Model at the most abstract level, through the Functional System Model, the Design Blueprint, and down to the code itself.

The maintenance process will "kick in" at the first model that will have to be changed as a result of the service request. A change in the nature of the business (which is fairly rare) will cause amendments to the Essential Business Model (process 9.2) and require a virtually complete development effort. While analyzing the client's new needs (process 9.1), it's more likely we'll discover that the new requirement necessitates a change in the *current implementation* of the system. If so, it's necessary to determine (process 9.3) if the changes needed are at the Functional System Model level (process

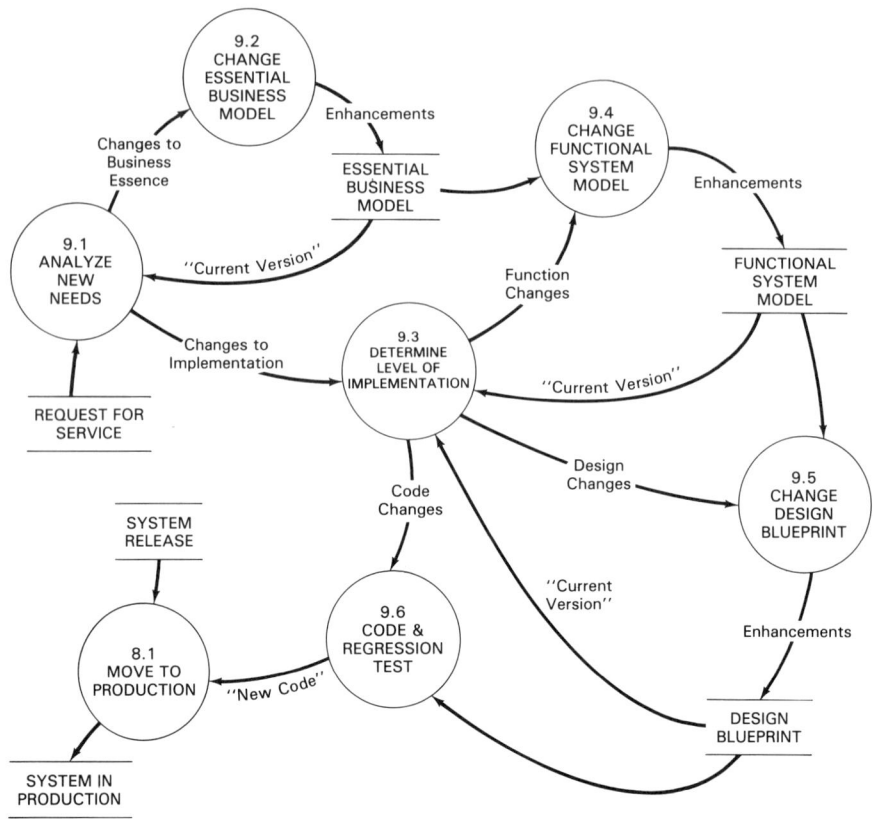

Figure 89: Maintain the System (9)—Version 1

9.4), at the Design Blueprint level (process 9.5), or the code level (process 9.6). While coding errors can be fixed at the code level with no other activity necessary, amendments at all other levels will require processes to amend other models of the system. And once the required changes have been made, the revised system must be moved to production. This process is exactly the same as 8.1, so 8.1 has been "reused" in this diagram.

Another approach to system maintenance is shown in Fig. 90. This model is exactly the same as the level 0 diagram of our SuperSet model with the exception of the numbering of the processes. The numbers of processes 1 through 8 of the original model are now prefixed with 9 to reflect that this diagram is a subset model of process 9 at the higher level, *Maintain the System*. The original process 9 is still number 9. The model of the final bubble in this diagram *is the entire diagram*, or, in other words, the context diagram for this diagram is bubble 9 in this diagram.

To carry this thought even further, this recursive definition of mainte-

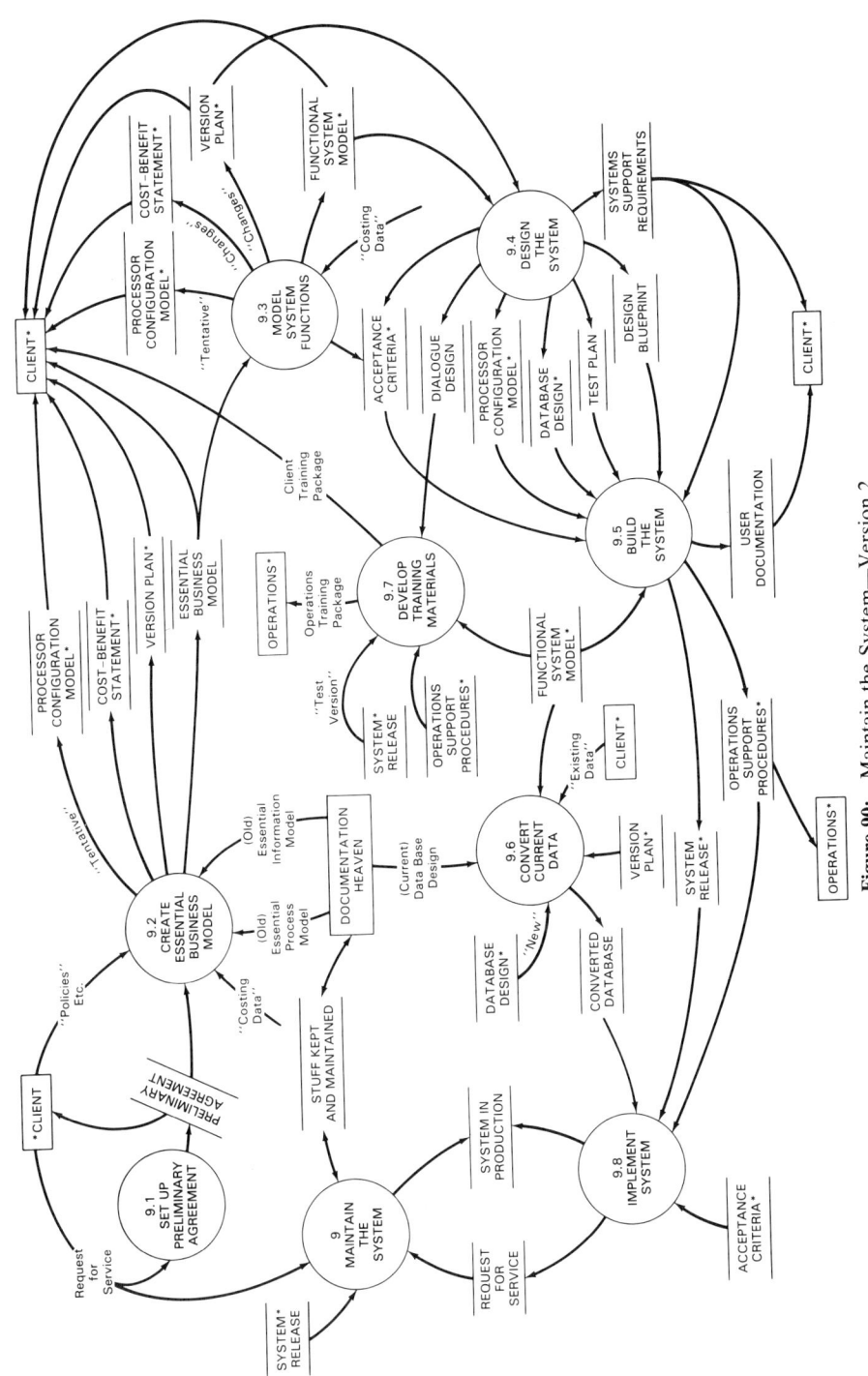

Figure 90: Maintain the System—Version 2

nance could also be applied to the level 0 diagram of the entire superset model by giving the *Maintain the System* process the number 0.

A Sample Real-World Model for Maintenance

In the preceding discussion we presented the abstract philosophy of how to handle systems maintenance and enhancement within the context of the superset approach. But the question is, How do we relate this directly to your real world of systems maintenance, where you and your staff spend most of your working life? What better way than to provide an example from a systems group within a financial services company that has already had to face this nagging issue. This particular group obviously has all the typical maintenance and documentation problems that face a mature organization and that accompanies millions of lines of old code. Again, what follows may not be totally suitable for your organization, since you will eventually have to build your own "maintenance model," but it shows that it can be a manageable process.

What Is the SDM for Maintenance?

SDM is an acronym for Systems Development Model. The SDM for Maintenance is a model for changing and enhancing a system to ensure its upkeep [Fig. 91]. The ideas and diagrams presented have been derived from a pre-publication manuscript by Trond Frantzen and Ken McEvoy entitled "A Game Plan for Systems Development".

How to Use This SDM

This is a model for implementing a change to a system in production. It isn't intended to be used verbatim, but should be modified with every new problem or request. (In large development or enhancement projects, we usually go through the process of producing a specially tailored model for the project to ensure we fully understand our approach to the problem. Due to the size of most maintenance requests, this is not possible or desirable.)

Therefore, the model should be used to trigger our thoughts when we start work on a request. By reviewing the model we can very quickly answer some basic questions such as:

- Do we need a cost estimate, or is approval not going to be dependent on the costs?
- Should we produce a written plan, or will it take longer to write the plan than to complete the changes?
- Did we update the documentation, both in our library and the user's procedures?
- Are we changing the functions of the system or are we correcting programming which didn't originally do what the user specified?
- Do we know how to test the changes which have been made?
- Do we have agreement with the user on what will constitute acceptable test results?

By answering these types of questions when we start a project, we will have a better chance of delivering the product that the user wanted.

This SDM is presented using the structured analysis tools as described by Tom DeMarco in his book called *Structured Analysis and System Specification*. The description consists of data flow diagrams, process specifications, and a data dictionary.

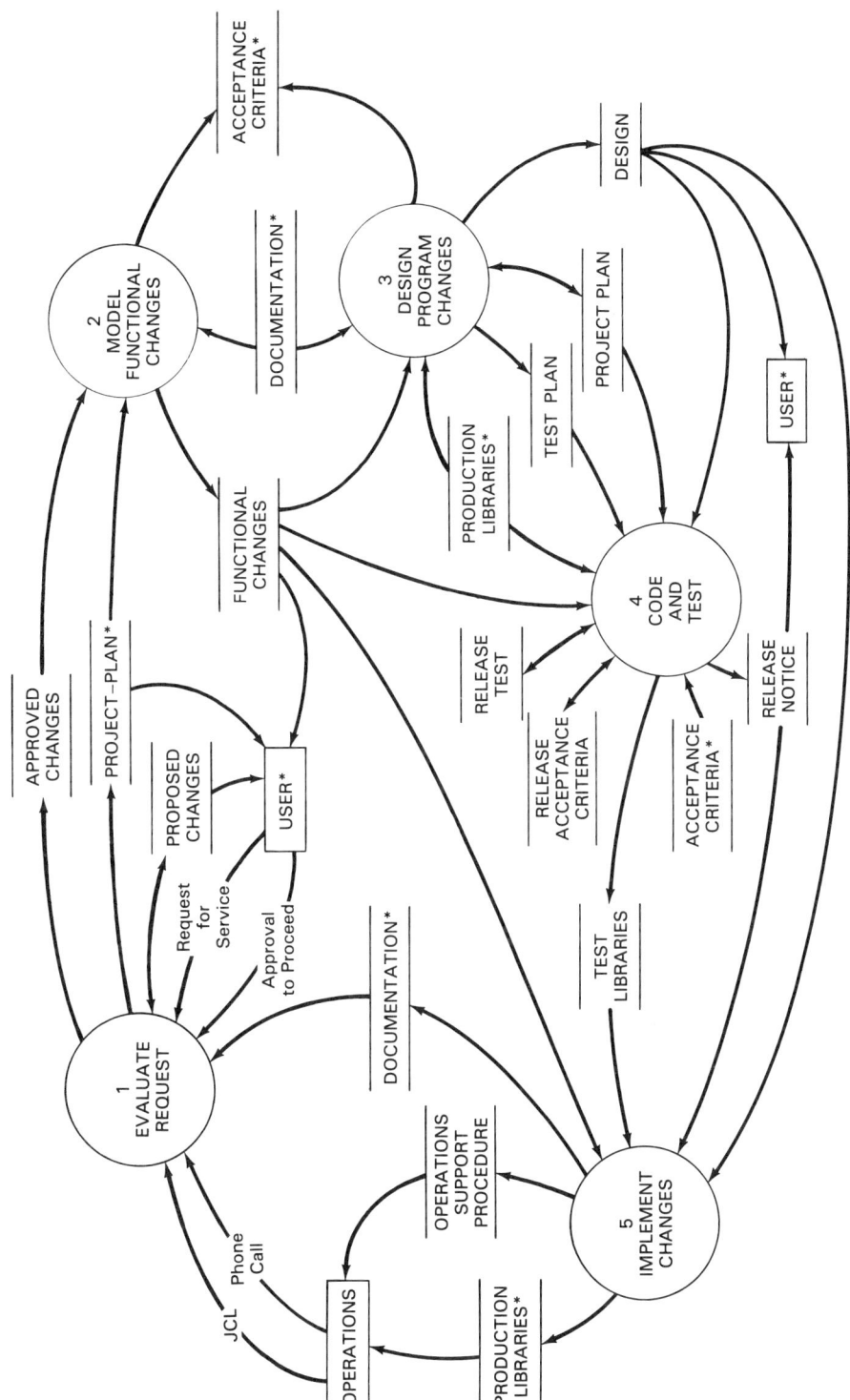

Figure 91: Diagram 0—SDM for Maintenance

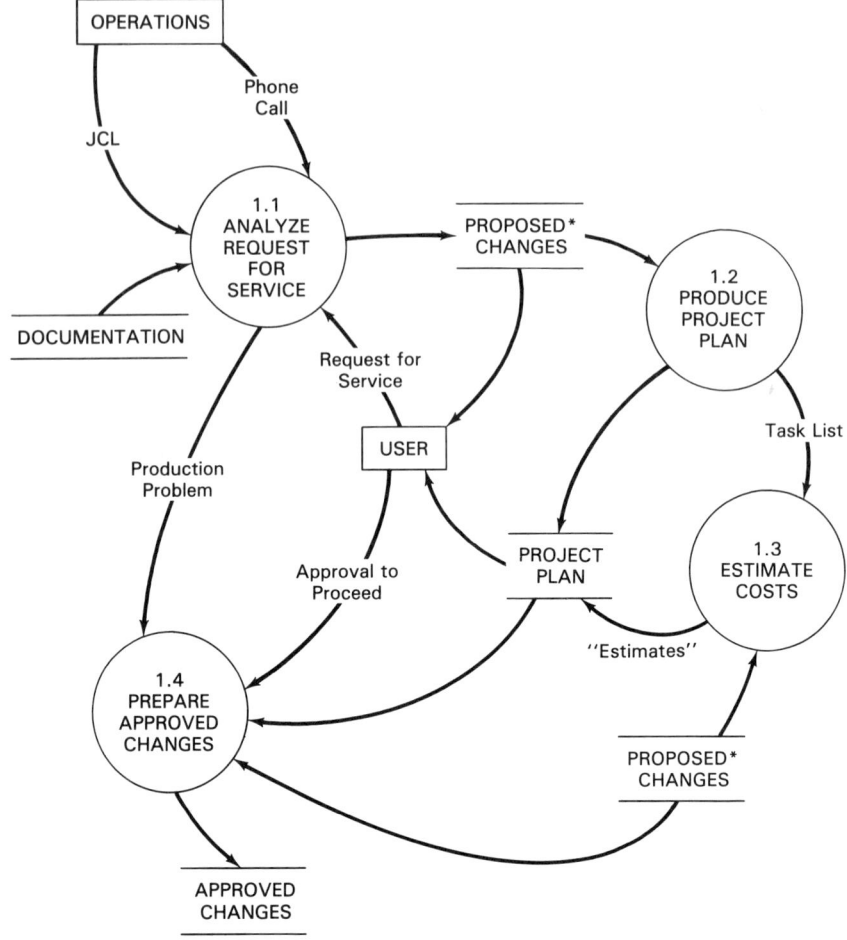

Figure 92: Diagram 1: Evaluate Request

1. EVALUATE REQUEST [Fig. 92]
    1.1 ANALYZE REQUEST FOR SERVICE
    - when a PHONE-CALL is received, get the JCL and determine if the request is a PRODUCTION-PROBLEM
    - when the JCL is delivered, determine if there is a PRODUCTION-PROBLEM
    - pass the PRODUCTION-PROBLEM on to 1.4 Get Approval to Proceed
    - when REQUEST-FOR-SERVICE has been received use the DOCUMENTATION to do a brief analysis of:
        a) changes to the system functions
        b) changes to programs/procedures
        c) the type of testing required
    - document your understanding of the problem and your recommended solution in the PROPOSED-CHANGES document
    1.2 PRODUCE PROJECT PLAN

Sec. 4.2   The Detailed Diagrams   **167**

- using the PROPOSED-CHANGES, break the project up into the tasks to be performed and write them onto a TASK-LIST
- estimate the time required to do each of the tasks (this is based on experience—your Supervisor or "buddy" will help you out at first)
- add a factor for unexpected changes (say 20–30%) [sic]
- plot this onto a work plan which is filed as part of the PROJECT-PLAN

1.3 ESTIMATE COSTS
- using the TASK-LIST, calculate the total number of days required to complete the tasks
- using the PROPOSED-CHANGES, estimate the machine costs
- add these up to produce COST-ESTIMATE, which is filed as part of your PROJECT-PLAN
- your users should determine the benefits of doing the changes and then give APPROVAL-TO-PROCEED.

1.4 PREPARE APPROVED CHANGES
- if a PRODUCTION-PROBLEM is received, approval is assumed (since the user probably needs the output) and we immediately proceed to the next step
- if not a PRODUCTION-PROBLEM, take the PROPOSED-CHANGE and add the key dates for completion and user involvement, and the cost estimates from the PROJECT-PLAN, and document it in a report or memo
- this document is the agreement of work which will be done in the rest of the project.

2. MODEL FUNCTIONAL CHANGES [Fig. 93]
    2.1 SPECIFY CHANGES TO FUNCTION
    - using DOCUMENTATION specify what functions the system should be doing differently
    - use data flow diagrams
    - walk-through with the user (you probably will do only one walkthrough of the functional change, so do process 2.2 as well before the walkthrough)

    2.2 SPECIFY CHANGES TO DATA
    - using DOCUMENTATION specify what information maintained by the system needs to be added or changed
    - walkthrough with the user

    2.3 RECORD ACCEPTANCE CRITERIA
    - using FUNCTIONAL-CHANGES, reformat into a list to indicate the criteria by which the changes will be accepted by the user
    - walk-through with the user

    2.4 REVISE PROJECT PLAN
    - using the additional information which is in FUNCTIONAL-CHANGES, determine if changes are needed to the PROJECT PLAN

3. DESIGN PROGRAM CHANGES [Fig. 94]
    3.1 DESIGN CHANGES TO DATA
    - using the FUNCTIONAL-CHANGES decide how the changed/new data should be stored (i.e., Copylibs, EZtables, Rate Disk, etc.)

    3.2 CREATE STRUCTURE CHARTS
    - using the FUNCTIONAL-CHANGES produce Structure Charts for the new or changed programs

    3.3 WRITE MODULE SPECS
    - write the minispec for each module which is to be created or modified
    - this should also contain specifications for changes to production procedures
    - walk-through with your Supervisor or "buddy" (for small program fixes; it should not be necessary to involve your user if he has agreed that the FUNCTIONAL-CHANGES are correct)

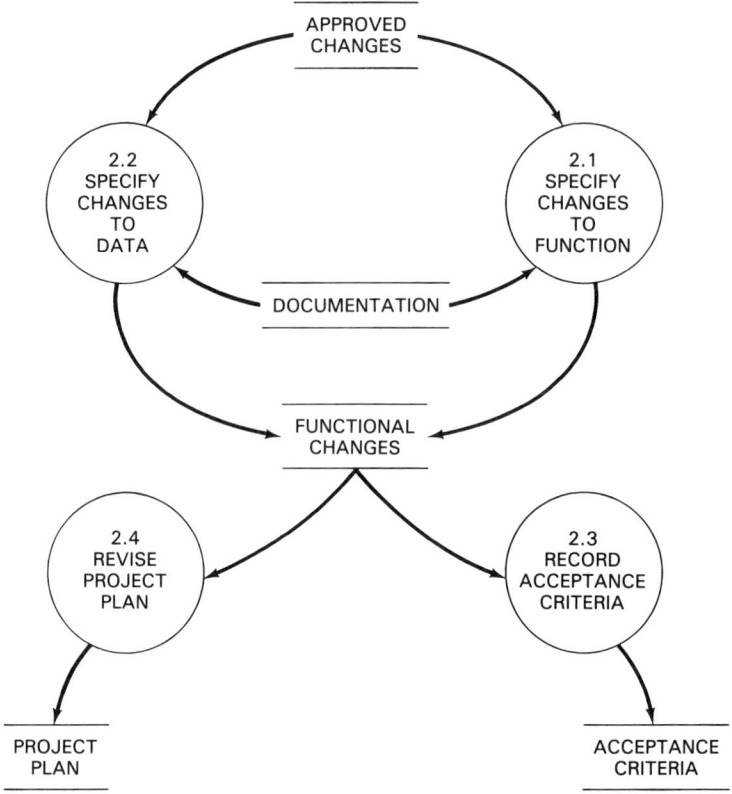

Figure 93: Diagram 2: Model Functional Changes

3.4 DESIGN CHANGES TO PRODUCTION PROCEDURES
  • using the existing procedures in PRODUCTION-LIBRARIES, specify all changes required due to the program and data changes specified
3.5 UPDATE ACCEPTANCE CRITERIA
  • add to the ACCEPTANCE-CRITERIA any physical requirements which must be met to ensure the acceptability of the system
3.6 DEVELOP TEST PLAN
  • using the DESIGN and FUNCTIONAL-CHANGES, produce a TEST PLAN (with your users)
3.7 UPDATE PROJECT PLAN
  • review the project plan in light of the design
  • make any revisions which may be necessary
  • (this process should continue throughout the project, but since it is always the same, it isn't shown in the rest of the model)
4. CODE AND TEST [Fig. 95]
  4.1 CODE
    • using the current programs in PRODUCTION-LIBRARIES and the mini spec from the DESIGN, code the program changes and put them into TEST-LIBRARIES
    • using the current copylibs, tables, etc. in PRODUCTION-LIBRARIES and the new/changed data in the DESIGN, make all necessary changes

Sec. 4.2   The Detailed Diagrams

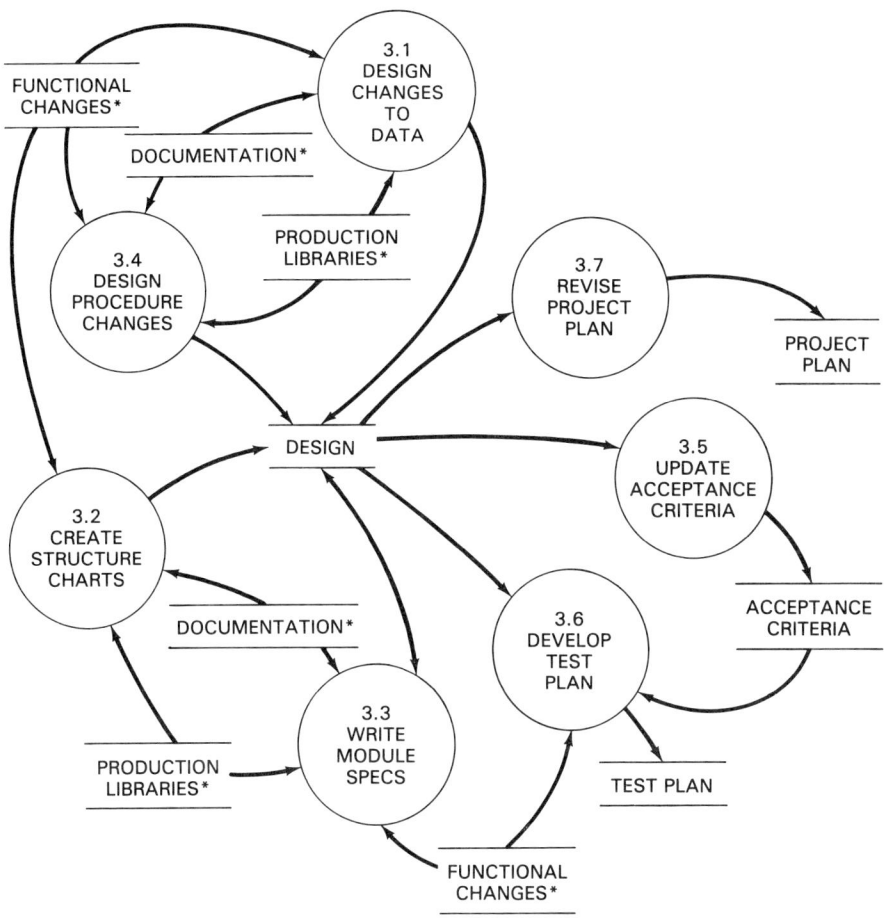

Figure 94:   Diagram 3: Design Program Changes

- walk through the coded programs prior to doing any testing (you could combine this with the walkthrough of the procedures which have been changed)
4.2 MAKE PROCEDURE CHANGES
- copy the current procedures from the PRODUCTION-LIBRARIES and make changes as specified in the DESIGN
- move them into TEST-LIBRARIES
- walk through the changes
4.3 BUILD TEST FILES
- produce TEST-FILES (perhaps from live data) according to the TEST-PLAN
4.4 TEST
- don't start testing until your walkthroughs have been done
- using the changed procedures and programs in TEST-LIBRARIES and TEST-FILES, run all the tests indicated in the TEST-PLAN
- measure acceptability according to the ACCEPTANCE-CRITERIA (the user should also review all test results)

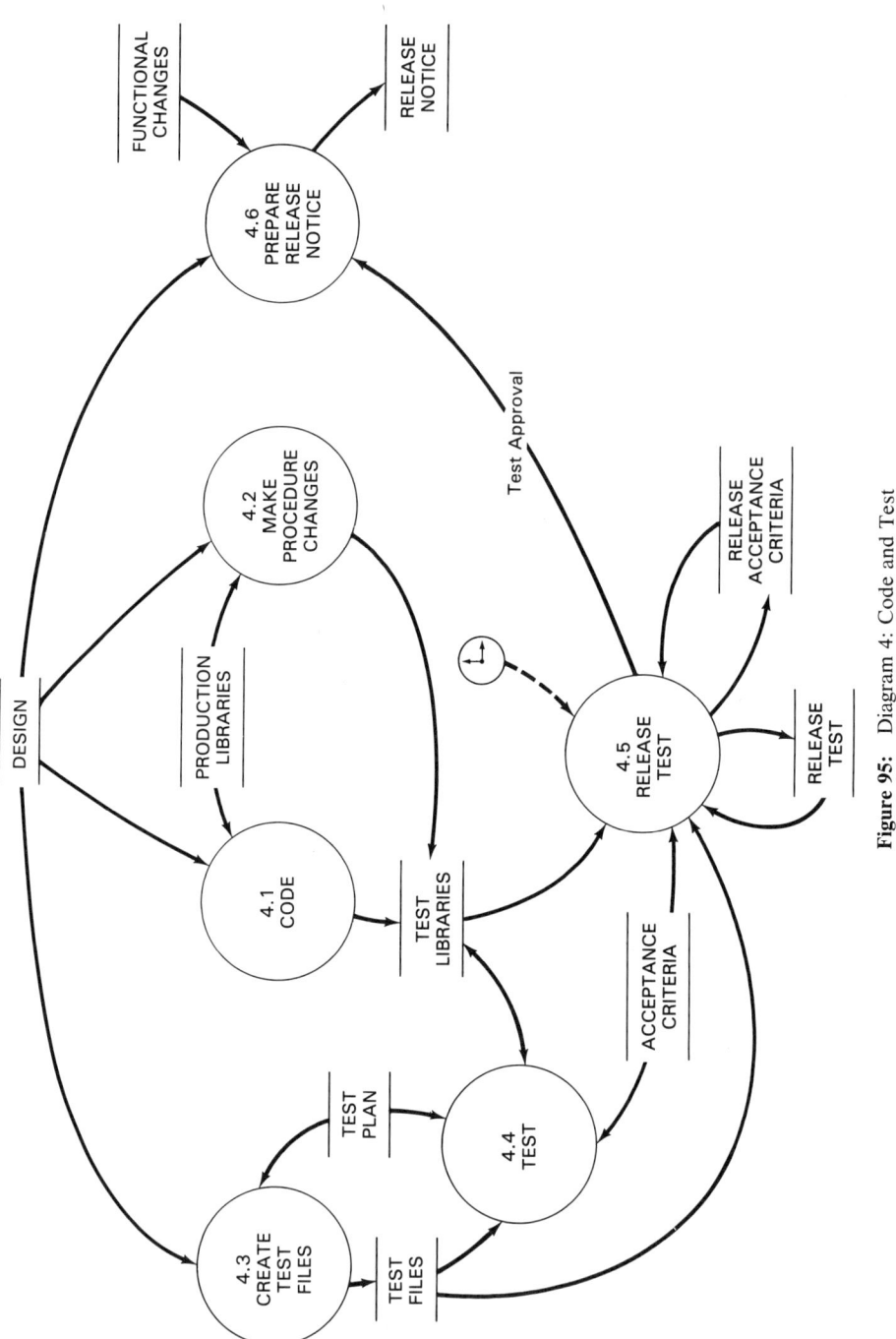

Figure 95: Diagram 4: Code and Test

Sec. 4.2   The Detailed Diagrams   171

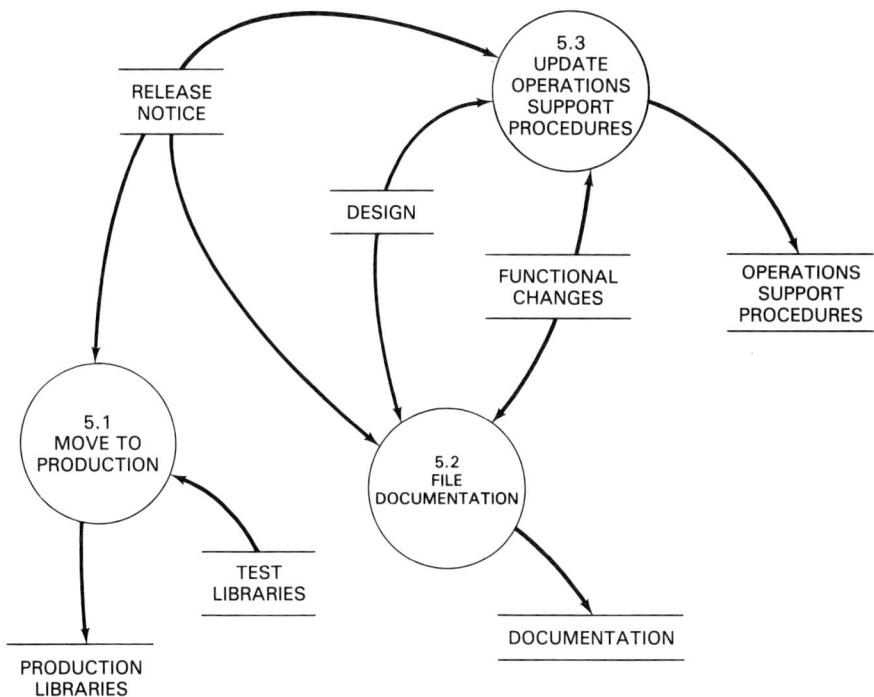

Figure 96:   Diagram 5: Implement Changes

   4.5 TEST SYSTEM RELEASE
- when it's time (perhaps monthly or quarterly) run the RELEASE-TEST
- also update the RELEASE-TEST with any new TEST-DATA from TEST-FILES
- when testing meets acceptance for release tests, produce TEST-APPROVAL

   4.6 PREPARE RELEASE NOTICE
- when TEST-APPROVAL is received, prepare the RELEASE-NOTICE according to the FUNCTIONAL-CHANGES and DESIGN

5. IMPLEMENT CHANGES [Fig. 96]
   5.1 MOVE TO PRODUCTION
- upon receipt of the RELEASE-NOTICE move the programs, copylibs, etc. from the TEST-LIBRARIES to PRODUCTION-LIBRARIES

   5.2 FILE ALL DOCUMENTATION
- upon receipt of the RELEASE-NOTICE file the FUNCTIONAL-CHANGES and DESIGN in DOCUMENTATION

   5.3 UPDATE OPERATIONS SUPPORT PROCEDURES
- upon receipt of the RELEASE-NOTICE update the OPERATION-SUPPORT-PROCS

DATA DICTIONARY

ACCEPTANCE-CRITERIA

> the criteria which the user will use to determine if the patch to the system was successful
> should not be anything new, but just a list derived from the FUNCTIONAL-CHANGES

APPROVED-CHANGES

> a memo or report which describes the work to be done in the project
> should be finalized only on approval from the user
> should include the following:
>   a. your understanding of the problem
>   b. the agreed solution
>   c. preliminary description of the changes required to functions, programs, procedures, etc.
>   d. dates which have an impact on the user
>   e. an anticipated delivery date
>   f. a cost estimate for doing the work
> distributed to:
>   a. the designated user
>   b. all others involved in the project, including other project teams affected, other support areas
> copies to:
>   a. user's supervisor
>   b. your supervisor

APPROVAL-TO-PROCEED

> acceptance from the user (usually in writing) of the PROPOSED-CHANGES, COST-ESTIMATE and PROJECT-PLAN
> gives us the authority to proceed

COST-ESTIMATE

> the estimate of the cost to work on a REQUEST-FOR-SERVICE
> includes both person and machine costs
> includes a budget for unexpected changes

DESIGN

> the specification for all changes to be made to the copylibs, programs, and procedures in order to accomodate FUNCTIONAL-CHANGES
> consists of updated Structure Charts, mini-specs, procedure changes, copylib and table changes, etc.

DOCUMENTATION

> all things which need to be kept and maintained and stored in DOCUMENTATION
Editor's note
This doesn't always exist for our current systems, and in some cases is not to be trusted. However, to resolve this we have specifically included a process in this model to allow for updates and filing of FUNCTIONAL-CHANGES and DESIGN in the DOCUMENTATION. By following this model, we will have reliable DOCUMENTATION in the future.

FUNCTIONAL-CHANGES

> a document which specifies the changes which are to be made to the system in production
> should be done using data flow diagrams and narratives
> changes to the data dictionary (if it exists) should also be specified in this document
> must be presented to the user for approval

JCL
- > supporting documentation for a PRODUCTION PROBLEM received by a PHONE-CALL, or
- > the JCL delivered in the morning which is scrutinized by the Program Librarians, who will detect a PRODUCTION-PROBLEM

OPERATIONS-SUPPORT-PROCS
- > the procedures manuals which contain all the instructions for the running of production jobs
- > run manuals which include job flowcharts, schedules, parameters, etc.

PHONE-CALL
- > notification from the Operations people of a PRODUCTION-PROBLEM
- > the phone call is followed up with a copy of JCL and other output which is used to Evaluate the Request

PRODUCTION-LIBRARIES
- > the libraries which hold all the production source and load programs, tables, and jobs which are to be run in production

PRODUCTION-PROBLEM
- > a special type of request which must be dealt with immediately, since it prevents the normal execution of a production run
- > all production problems have prior approval

PROJECT-PLAN
- > the plan which is developed for the project based on the tasks which must be performed
- > the Keane project management methodology and forms should be used

PROPOSED-CHANGES
- > the result of the preliminary analysis of the problem
- > could be in the form of a memo, but for smaller patches it will be a discussion or phone call
- > should include the following
    a. your designated user and contacts in other systems areas if required
    b. your understanding of the problem
    c. a proposed solution
    d. estimates of the changes required to functions, programs, procedures, etc.
    e. a guesstimate of the likely cost of doing the work (this should only be provided if the user demands it and only upon approval by your supervisor)

RELEASE-NOTICE
- > a memo which explains the contents of a systems release
- > based on FUNCTIONAL-CHANGES
- > explains to the user the impact of the changes
- > (should be nothing new to the user, since he has already been walked through the FUNCTIONAL-CHANGES)

RELEASE-TEST
- > a specialized volume test which is run on a regular basis to determine the acceptability of changes prior to a move to production

> each system will have its own special release test and acceptance criteria

REQUEST-FOR-SERVICE
> any user-initiated request for a change to the system

TEST-LIBRARIES
> the place where test programs are stored prior to a move to production
> all testing is done from this library
> contains programs, procedures, copylibs, tables, etc.

TEST-APPROVAL
> a notification to the process "Prepare Release Notice" that the release test found no problems with the programs in the TEST-LIBRARIES which need to be moved to production

TEST-FILES
> files of data used to test programs in the TEST-LIBRARIES

TEST-PLAN
> the description of the procedures to be used to test the programs, tables, and procedures in the TEST-LIBRARIES
> the ACCEPTANCE-CRITERIA and FUNCTIONAL-CHANGES are used to produce the plan

In MaxiMoney, Inc's,[27] model for maintenance it's debatable whether some of the data dictionary entries should really be so descriptive of the related processes. There's also a considerable measure of redundancy between the data dictionary and the process specification, which was rationalized as "controlled redundancy." But that's not the point. For this company, for its purposes, for its people, it describes clearly what's wanted, and it works for them. In whatever manner you may ultimately define your model of the maintenance process, whatever works for you is right—as long as you think it out and model it, and don't leave it up to the maintenance people to guess what the procedure should be.

[27] With special thanks to Rommie Vanderboor, a DP director at MaxiMoney, Inc. MaxiMoney, Inc., is a pseudonym for one of North America's larger insurance, financial services, and real estate investment firms. In keeping with its usual desire to keep a low profile, the company prefers to remain anonymous. However, if a reader has a serious desire to speak with the director responsible for successfully implementing the superset approach and the structured techniques at MaxiMoney, Inc., please contact the authors at the address on the response form at the back of this book.

CHAPTER 5

IMPLEMENTING A SUPERSET APPROACH

5.1 INTRODUCTION

Although the main thrust of this book is a discussion of the SuperSet System Development Model, its working title is *A Game Plan for Systems Development*. As a game plan we must discuss some aspects of systems development that are outside the confines of the SuperSet System Development Model but are certainly part of a systems development methodology. One of those issues is how we can implement the superset approach in your shop, which we'll discuss next. Another is effective training of systems developers discussed in Chapter 6. "Manager's Notebook," Chapter 7, reviews some ways of doing better work faster.

5.2 TRAIN USERS IN THE SUPERSET MODEL CONCEPTS

For the same reasons we should train systems development people in the concepts and methodologies inherent in the SuperSet System Development Model, we should also train the users.

To participate in the systems development process as leaders or developers, users need to understand the process of developing and managing the development of systems. One way to achieve this is to teach them the concepts contained in this superset. A thorough understanding of the basic development process—what's involved, why it seems to take so long, why there are so many deliverables, why systems development must proceed in a certain organized manner, and how it can be fast and dynamic—will give users an insight they may not have had before the training sessions were held. Their improved understanding of the generic development cycle will enhance and promote the Development Team's productivity.

A superset training program will also bring users into the mainstream of systems development, where they should be. Being more involved, they will be less likely to criticize the data processing staff and the difficulties DP has always faced. By understanding the process and being involved on a day-to-day basis, the user will also have a greater sense of responsibility and ownership of the delivered product.

Although the various methodologies may change with time, the concepts forming the base of the superset are fundamental and will stay with us for many years. Because user analysts will be members of Development Teams, they and their management must have a thorough understanding of the manner in which the SuperSet model should be used. This understanding on their part will enhance productivity and promote quality.

The training program you develop will naturally depend on the goals you want to achieve.

5.3 HOW IT WAS DONE AT MAXIMONEY, INC.

MaxiMoney, Inc., with its head office in Toronto, Canada has long been committed to advanced systems technology, and certainly to the ongoing improvement of the technological and business skills of its data processing professionals.

As a matter of background, the firm is a large insurance, financial services, and real estate investment company with assets of more than $20 billion. Although the data processing function is decentralized, it has more than 500 systems professionals in head office. The company is international in scope. It does about 60 percent of its business in the United States and is also a well-known institution in the United Kingdom.

A Commitment to Excellence

MaxiMoney, Inc., is sufficiently committed to staff quality and advancement that it has its own data processing education sections. The company has had a skeleton systems development model (SDM) since 1972. Structured programming has been used since the early 1970s; structured design, especially the data-structured methodologies of Jean-Dominique Warnier, Ken Orr, and Michael Jackson, has been popular since the late

Sec. 5.3 How It Was Done at MaxiMoney, Inc.

1970s, and more recently that of Larry Constantine as described by Meilir Page-Jones. Structured analysis, based on Tom DeMarco's work, was formally adopted in September 1981. The Keane project management methodology has been in use since 1980. Walkthroughs and inspections of code have been common practice for several years, and the autumn of 1982 saw the formalization of walkthrough guidelines for analysis and design deliverables. In 1983 several data processing departments within the company embarked on an aggressive program to use information modeling techniques. One such department, using the techniques and deliverables identified in this book on the SuperSet System Development Model, started *integrating* Information Models and Process Models in an attempt to improve the quality of deliverables and produce more stable systems.[1]

Successful implementation of the concepts and methodologies behind the SuperSet System Development Model requires a philosophical commitment to people and quality through education and support. It also requires that the commitment to excellence extend to the pocketbook!

A few companies, in their pursuit of excellence, simply send some of their managers and supervisors on three-day "Learn Everything You Can About Systems Development" courses, expecting them to know all there is to know, and then to populate instantly all their staff with this newfound knowledge. Instant expertise for $700.

Unfortunately, the learning process just doesn't work very well this way, although it is a great start. It takes time to acquire skills in the systems building business, just as much as it does in medicine or law or engineering. It also takes the commitment to establish a budget and an apprenticeship program. Knowledge and expertise do not come cheaply, and any company that wants to be on the leading edge had better make sure its people are as advanced as the technology.

MaxiMoney, employing the principles of the SuperSet model, is certainly one company that has made that commitment. Not only do data processing people regularly attend external seminars and courses, many courses are purchased or developed and then offered internally on a regular basis. And professional full-time consultants provide continuing assistance and expertise through an application development center concept, or by being on long-term loan to departments wishing to develop improved systems development strategies. This is all in keeping with Thomas Carlyle's conviction that "we need reminding as much as we need educating." This

---

[1] MaxiMoney, Inc., is a pseudonym for one of North America's larger insurance, financial services, and real estate investment firms. In keeping with its usual desire to keep a low profile, the company prefers to remain anonymous. However, if a reader has a serious desire to speak with the director responsible for successfully implementing the superset approach and the structured techniques at MaxiMoney, Inc., please contact the authors at the address on the response form at the back of this book.

company clearly understands that there is more to learning and quality than just taking a course every now and then.

One of the company's marketing vice presidents, in a meeting with several data processing directors, said: "We are going to need a much better educated system developer in the future; a system developer will have to have the experience to know when to use what tools, when to push what processes, and how hard." It was agreed that a basic education program was in order.

The Basic Training Program

The following program was developed and instituted. Perhaps you can initiate a similar program for your company.

- First, many of the principles and methodologies now contained in this book were accepted by several directors and managers (but not all of them) as fundamentally desirable.
- Second, a short course and videotape specifically about the understanding and use of the SuperSet model and its essential deliverables was produced. The training objectives were simple and straightforward:
  - understand the superset concept;
  - know the nine activities of the life cycle;
  - know the six key deliverables and which of the nine activities they result from;
  - know where the key management checkpoints should be;
  - know how to use the SuperSet model's documentation and its Dictionary.

The basic education program was then offered to company staff who had an interest in developing systems. The "Sesame Street" approach was used in both the video and the instructor's tutorial (i.e., repetition, repetition, repetition). The tone of the course was changed easily by the instructor when class makeup changed from users to project leaders, managers, and supervisors, or to systems analysts and designers. Ultimately, the course and video were given to about 350 professionals at all levels, including about fifty senior level line area users.

- Third, the Professional Development and Research department had the mandate to constantly monitor, maintain, and enhance the base SuperSet model and monitor the methodologies available to systems development professionals.
- Fourth, a team of "experts" would provide consulting services to user groups concerning the SuperSet model and the various development methodologies.

This was the basic program that introduced systems professionals and users to the concept of the SuperSet model and a consistent set of tools and techniques—a game plan for systems development. But there was more. There had to be a distinct training program for analysts, designers, project managers, and users in the specific tools used to develop new systems. To ensure that new methods were not applied with old habits, some rethinking was called for. We discuss basic staff training in Chapter 6.

5.4 MANAGEMENT COMMITMENT—HOW MUCH?

The use of this game plan for developing systems and all the methodologies implied by it is not something systems staff can readily implement department-wide on their own.

First, to do it in a methodical way calls for a budget. Education, knowledge, skills, and leading-edge quality systems don't come cheaply. Your belief that building new business and technological systems is an excellent investment and that it will result in very substantial returns to your firm is fundamental. You won't always be able to prove this using conventional cost-benefit techniques. But all you have to do is look at what your competitors are doing, use your imagination, and believe that you too can accomplish as much or more with people and systems. And then devise the strength of argument and conviction to get the money to upgrade the skills of your people.

Second, to do it successfully requires the total support and dedication of purpose of senior company management.

Third, you will need the enthusiastic support of the people doing the work—and with consistency and an identifiable program, this will be easy to come by.

Fourth, and vitally important, there has to be the opportunity for constant dialogue with people having a thorough understanding of the SuperSet model, the concept of flexibility that is its foundation ("adapt, don't adopt!"), and the various methodologies needed to physically build systems.

Without these four elements it is unlikely that any implementation effort will be maintained.

5.5 EMPHASIZE ANALYSIS AND DESIGN

Few organizations will effect real change without also making it clear to all staff that the fundamental working philosophy has changed.

It hardly helps to promote the improvement of front-end analysis when the DP manager continues to hire the generic programmer/analyst. While programming is a very important function in the development process, the programming discipline will not provide the business solution or devise the architecture needed to create a system. Management of the organization must now begin to emphasize analysis and design, not programming. Programming is not part of the problem any more, it is an effect.

Emphasis on analysis and design will get the DP professionals out of their programming cubicles and into the user's world, doing things that will have much more leverage for us. The IEEE did a study that included IBM, GTE, and TRW, and found that 76 percent of all problems were not problems of program design, implementation, or coding; they were oversights and flaws in the analysis and systems design process.

The solution to this problem would appear to be to emphasize analysis and design. Of course this also implies an education program for new analysts and designers, and perhaps a reeducation program for programmers who are about to be promoted to the analysis and design arena.

5.6 MAINTAIN THE SUPERSET MODEL

The systems development process is as dynamic as any other process in your business, and your SuperSet Systems Development Model must be maintained and enhanced to reflect changes in your approach to the development of systems.

These changes could result from changes to the essential business activity of your organization, but more frequently they will result from changes in the operating environment. The latter could be such very physical changes as new hardware or development languages, and some less tangible changes such as new analysis and design tools (yes, these will change in time, too!), or changes in government regulation.

Some group in every organization must be charged with the responsibility of maintaining the SuperSet Systems Development Model; adding new processes that have been observed to be necessary in the creation of systems, and deleting processes made obsolete through change. The mentors, or equivalent group of senior analysts and designers, will be the ones to notice the needed changes, but there must be a management commitment to gathering these observations and acting on them.

A SuperSet model that is not dynamic, that is not maintained, and that does not grow with the development of the staff and changes in the business is no more useful than any other off-the-shelf methodology. In fact, it is dangerous because it freezes us at a given level of productivity and exposes us to challenges from more creative competitors. A "we've always done it this way" approach to systems development will always lead to high cost, and inflexible systems.

5.7 IMPLEMENTATION PAYBACKS

If you implement the concept of the SuperSet System Development Model and the methodologies suggested by it, the payback to the data processing department, the users, and the company will be very substantial. In a nutshell, you can expect the following:

- users will understand their systems and take greater responsibility for them;
- data processing professionals will understand their users' systems;
- people will become trained as analysts;
- there will be a broader base and more thorough knowledge of corporate systems;
- documentation will be up to date and of high quality;
- resource commitment to "maintenance" will be reduced, while new development activities can increase;
- future front-end analysis effort will be substantially reduced because the Essential Business models invested in today will be available to tomorrow's developers;
- fewer people will be needed to do more and better systems development work in the future.

Almost one hundred years ago John Ruskin wrote:

> It's unwise to pay too much, but it's worse to pay too little. When you pay too much you lose a little money, that's all. When you pay too little, you sometimes lose everything, because the thing you bought was incapable of doing the thing it was bought to do. The common law of business balance prohibits paying a little and getting a lot. It can't be done. If you deal with the lowest bidder, it is well to add something for the risk you run; and if you do that, you will have enough for something better.

5.8 PROTOTYPING AND THE SUPERSET MODEL

A lot of questions are currently being asked about prototyping. How does it fit into our methods of developing business systems? How does it fit into the regular systems life cycle? What tools should be used in conjunction with prototyping? Do analysis and design still have to be done?

As with any project, the creation of a unique life cycle for the new system involves studying the SuperSet System Development Model and asking for every activity:

"Must this be done?"

The risk of *not* doing each activity is assessed, and if it is deemed low enough the risk is accepted and the activity is not included in the specific development model for the target system. In this sense, prototyping fits in very nicely with the SuperSet model, because prototyping is just another strategy, a strategy made possible by different and newer tools. However, the questions that must be answered are:

"What is the risk of adopting a prototyping strategy; and how do you measure that risk?"

"What systems life cycle activities can be dropped (or scaled down) if a prototyping strategy is being used?"

Prototyping: Pros and Cons

Prototyping can be defined as the development of a working model of a system, with no frills, and with the intention of using little or none of the prototype code in the final implementation.

On some projects the use of prototyping tools can result in some functions being delivered to the user very quickly. Also, the user gets to work with real output and screens early on, without being locked into anything. Changes are usually quickly implemented. Yet, from another point of view, prototyping is "rushing to code," and it's coding too early that has caused so many problems in the past.

In order to be a candidate for prototyping, a project should probably have all of the following three features:

1. *Small*: Problems developed in the past when there were attempts to write code before the requirements were understood. The smaller the task, the greater the chance that the user's needs are fully understood.
2. *Available Data*: To be able to prototype a project (which is clearly different from building *scenarios* or *versioning*), we have to assume that the database already exists in a reasonably accessible form, and the developer only has to do the necessary work to form her or his own view. Without accessible data, good prototyping tools are as useless as a car without gas. And if you don't have the existing data in a normalized form, we must create an Information Model first. We can't prototype an Information Model; that's dangerous.
3. *User Implemented*: If the project is being implemented directly by user staff, rather than by an external data processing staff or consultants

Sec. 5.8 Prototyping and the SuperSet Model

(although they can certainly advise), the people doing the work will already have an "internal model" of the system, even if no effort has been made to put it on paper. If the users are doing the work, it is hoped (but not guaranteed) that they will understand the requirements.

Also, the prototyping tools must be ones that can be used in a production environment. Alternatively, there must be a commitment by users to the conversion from a prototyped environment to a production environment. Many companies are now suffering under the tyranny of unmaintainable systems in APL or BASIC that were originally seen as prototypes.

With larger projects the actual process of coding will usually be less than 20 percent of the total project time. Even a fourth generation language that doubled our coding productivity would only produce a 10 percent saving over the entire project. This is good, but not revolutionary. It is still absolutely mandatory that the essential business requirements, an information model, and complete user requirements be fully stated and understood before considering physical implementations. At what stage any systems model is considered sufficiently complete to start prototyping is an issue of risk assessment and tactics to be determined by each project manager.

Previously, in section 4.2.2 under the subsection *Productivity and the Essential Business Model*, we discussed *event-based analysis* and *essential modeling*. We also illustrated how an entity-relationship diagram and a high-level Essential Process Model could be created rapidly to reflect the innermost substance of the business requirements. Event-based analysis is crucial to successful prototyping.

To prototype effectively, it is essential that the following be developed, in this order:

1. Entity-Relationship Diagram
2. Event List
3. Event mini-models
4. Integrated Essential model of the system (extrapolated from item 3 above)
5. Logical database design (normalized)
6. Physical database design

Basically, you can then go ahead and create prototypes safely and quickly within the context of the overall design, with a warm 'n fuzzy feeling knowing that the business essentials have been covered. Although database design cannot be prototyped (an opinion strongly held by the authors, based on disastrous experience), it—by necessity—can be *iterated* based on prototype scenarios. How much iteration is a function of the particular DBMS used, and the nature of the dialogues required.

Bernard H. Boar has written an excellent book on this subject entitled *Application Prototyping*.[2]

Prototyping and the SuperSet System Development Model—and the structured development techniques—don't seem very incompatible after all. The structured techniques were the original form of prototyping (top-down, iteration, refinement, letting the user see something, etc.). Both prototyping and the structured techniques are concerned with maximizing communication and user involvement.

---

[2] Bernard H. Boar, *Application Prototyping* (New York: John Wiley & Sons, 1984).

CHAPTER 6

EFFECTIVE SYSTEMS TRAINING

6.1 OBJECTIVES OF TRAINING

It's generally agreed that analyst and designer skills are vastly different, although complementary, and these should be separate career paths. Should be. But reality dictates that this just isn't possible.

Many systems development managers also recognize that there are too many generic "programmer/analysts" trying to build new systems, with few specific analysis or design skills. Since most analysis and design education programs seem to be developed by the same people who created Upside Down World, much of what systems developers do is seen as black magic or an obscure art learned by osmosis in a back room from rarely seen gurus. But with technological change affecting the systems engineering business as much as it is traumatizing workers in the steel industry, we just can't afford to rely on good intentions any more. To become or simply remain competitive, we need to develop a pool of skilled, competent analysts and designers as quickly as possible.

The traditional career path begins with entry level programmers progressing to be designers and, later on, analysts. This means that entry level people work as programmers on project A. When they are deemed ready for broader experience, they work on design with project B. Later they do analysis with project C.

However, the systems development life cycle (SDLC) doesn't work that way. Analysis comes first (at least, usually), followed by design, then programming. Designers are doing a job that is analysis-driven, but quite often they know nothing of analysis. Design and analysis are quite intimately connected. Designers need to know something about analysis, and analysts need to know something about design.

With these principles in mind, we suggest a strategic plan to train analysts and designers in a manner that is in harmony with the SDLC. The idea is to present first things first and teach people the fundamentals of each discipline. This will also enable them to make intelligent career choices, rather than just falling into the discipline they start out with.

Both groups need two basic skills:

1. *information modeling*: to define the data resource in terms of entities and their relationships, and provide the data dictionary for the overall systems model:
2. *process modeling*: to define the business processes (what users want done), possibly using data flow diagrams and text.

These are both top-down disciplines that produce models of the real-world subject matter. They help define underlying business policies. They complement one another. A complete systems model needs both processes and data, just as a complete sentence needs both verbs and nouns. Information modeling is the noun; process modeling is the verb.

To train people in these skills requires a combination of formal education and experience on the job. The timing of this training is crucial. Candidates should take the appropriate courses shortly before starting work on a project where they will use what they have just learned. Then they should work in an "apprenticeship program" under the guidance of an experienced analyst or designer. (Remember that this procedure does not ensure competence; it only guarantees that the person has had training and experience in the particular subject area.)

When staffing projects under such a program there are a couple of points to keep in mind:

- process modeling and information modeling skills must be represented on every development team;
- nobody should do the work without having formal training first;
- don't assign the "first available body" if that body lacks the needed expertise; if nobody has the needed expertise, hire a consultant;
- there should be no forced assignments to either analysis or design; that

is, management should not impose a career path on staff; let people do what they think they like the most.

6.2 CORE TRAINING

The nature of the development life cycle makes it possible for analysts and designers to gain hands-on experience in a real situation with the theory they learn in courses.

This education strategy is intended to break down the barrier between data processing and user areas and increase their mutual understanding. It consists of formal courses identified as either "core" or "recommended" for analysts, designers, and management.

For example, DP analysts and user analysts have the same course of study with one difference: structured design is recommended for DP (system) analysts and is not applicable to user (business) analysts. This is because DP analysts must have some knowledge of how to implement functions inside the automation boundary of a system, while user (business) analysts need only be concerned with the functions themselves.

Highlights for DP Analysts and Designers

Unless new hires are fully experienced in both process modeling and information modeling, they should also take the core training courses. The specific courses recommended are generic and available under various names in the general marketplace, but were chosen because of a desire to standardize methodologies.

1. *Basic Training Required*
   Basic programmer training: rookies take this; it's also pretty well unique to every company.
   An internal systems development model (SDM) course: this also will exist in your company, or needs to be developed (possibly using the SuperSet model as a base).
   Sufficient programming experience, especially for aspiring designers; each company will have to define what "sufficient" means.
   Participation as junior members of a development team doing models of systems.
2. *Structured Analysis Training*
   Analysts need to do it; designers need to understand it. This training will teach how to develop the Process Model and Data Dictionary.
3. *Information Modeling Training*
   Analysts and designers need it. This teaches how to develop entity-relationship diagrams and expands on development of the Data Dictionary.

4. *Structured Design Training*
Designers need it; analysts don't. This teaches how to develop design specifications and the Design Blueprint.

6.3 TRAINING FOR MANAGEMENT

In this context, there are two "management" groups. Regardless of actual title, the functional responsibilities are these:

1. *"Primary Managers"* are actively engaged in the day-to-day supervision of systems development work. Therefore, to do an effective job, they must understand what their people are doing and why they are doing it—that is, they themselves must know how to do the work. They must also develop appropriate project management skills.
2. *"Secondary Managers"* are one or more levels removed from the nitty-gritty activities of systems development. They are usually, but not always, management level people concerned with broader issues. They need general familiarity with concepts and techniques rather than detailed how-to knowledge.

6.4 TRAINING FOR USERS

User analysts are people from user areas who participate actively in the analysis phase of systems development. They may have specialist or management jobs in their respective departments, and they bring essential knowledge of the business to the development team. Unlike their DP counterparts, user analysts do not need to fully understand or participate in the technical aspects of design.

User management in this context is the user with the systems development budget. She pays the piper and to some extent calls the tune. However, she is at least one level of management away from the project's "doers." Like the secondary manager, the user manager needs to understand concepts of analysis.

6.5 TIMELINESS OF TRAINING

Every member of a development team who does not have the basic required education or skill level needed by a project must attend the right courses to learn the right theory. The person's major hands-on education comes when he or she is working on a project.

In many companies, especially larger ones, it's a fairly common practice to send people on analysis and design courses "between projects" or "when there's time." Although this makes scheduling and possibly even budgeting relatively easy, it makes real education almost impossible.

Students generally invest four or five years of their youth at a university learning about their chosen discipline. This is the theory part of their education. When they are finished there are "commencement exercises." Some students consider this to be the end of their days of studies and will probably never open a book again. But many others understand that their graduation is the beginning, not the end, and that's why it's called "commencement." They know that when they enter the real world, they will acquire practical ways of implementing what they have learned. The theories they have studied for several years will then be put into real business situations which make them almost understandable. There will be a lot of "ah-ha's!" where even the best teachers couldn't eliminate the dense fog factor. There will be a time of experience, and through that experience there will be exponential growth in knowledge and understanding.

In systems development, however, many otherwise responsible managers expect novices to learn the theory of analysis and design (or a new methodology) and to apply that theory with reasonable success without ever gaining any hands-on experience or working directly with a mentor. This concept is something like learning neurosurgery by correspondence. The theory is fine, but the application leaves something to be desired.

To overcome this problem there must be a well-defined education program, but also a qualified mentor. The scheduling of that education must also be well planned. If we can agree that it is senseless to learn ADS/O in January but not get the opportunity to apply it until September, then it should be equally senseless to send people on analysis or design courses "when there's time." We don't inflict this kind of pain on programmers; why should we do it to analysts and designers.

Timeliness should mean swift application of the theory learned, with expert help. And theory, if at all possible, should be acquired within two weeks of working on a project to apply that theory. The end of a training course should be "commencement"; and the application should be immediate.

6.6 QUESTIONS OF STANDARDIZATION

Standardization of analysis and design does not mean rigid adherence to a set of rules, but an agreement to use the same tools so that we can communicate with one another.

There are many ways to do structured analysis and design; their notations are different but all of the different methods are fundamentally sound. However, for both notation and education, as well as innovation, we suggest you commit your company or department to one of the more recognized "schools" of analysis and design. There are four reasons for this:

1. All of the recognized schools of structured analysis and design provide *consistency of techniques*, including some form of information modeling.
2. All of the recognized schools of structured analysis and design provide *methodological links* across the different development disciplines.
3. All of these schools transcend business and cultural boundaries; their literature and consultants, *all speaking the same basic language of methodology*, can apply the structured techniques to any business sector.
4. The recognized schools of thought are veritable spawning grounds of *innovative* individuals who continually bring the art and science of systems development further into the twenty-first century (admittedly, sometimes kicking and screaming).

6.7 CORE FORMAL EDUCATION

Figure 97 does not suggest that education should be limited to just these few courses. The systems development business is far too complex to be learned from such a short selection. It is, however, the essential education needed to learn the theory behind the concept of systems development.

C = Core material R = Strongly recommended	Systems Analysts	User Analysts	Systems Designers	Primary Managers	Secondary Manager	User Management
SuperSet System Development Model	C	C	C	C	C	C
Structured Analysis	C	C	C	C	R	R
Information Modeling	C	C	C	C	R	R
Structured Design	R		C	C		
Project Planning & Control	R	R	R	C		

Figure 97: Core Formal Education

6.8 SKILLS INVENTORIES

Developing systems requires a broad spectrum of skills and a considerable depth in each. It's sometimes difficult to keep track of who has what skill or who needs additional training to enhance their knowledge. Few people have or can have all the expertise across the range of skills needed.

One way of overcoming the problem of keeping track of staff knowledge is to develop a "resource inventory" spreadsheet of the skills found most important to the development of quality systems and to the improvement of productivity in your company or department.

Simply list the people across the top and the essential skills along the side of the spreadsheet. For each, indicate "M" if the person is a *main* resource for the skill, "S" if a *secondary* resource, and "P" if the person has the *potential* to become a resource for the skill. Skills should include technical as well as business knowledge of specific systems. Review the spreadsheet regularly to add or delete items. Refer to it when a termination occurs or an important change is imminent. Concentrate on turning the P's into S's and M's.

Doing this helps us:

- recognize and relieve an overburdened resource;
- enhance and enrich an underutilized resource;
- rearrange resources coherently when an opening arises;
- identify which employees (regardless of position) really do, or do not, contribute to the team; and
- communicate to subordinates the skills needed and who provides them.

Share this spreadsheet and its updates with other managers or project leaders. The available resources, by specific skill mix, can then be called upon by other managers for some fast-track consulting or even project participation.

This does not imply that a "pool" of resources should be created. Teams should be encouraged and kept reasonably intact.

6.9 LEARNING THROUGH MENTORS

In our quest for improved productivity and more knowledge, we have identified the need to develop the skills of our people by formal training and experience on the job—an apprenticeship program. Obviously, such a program needs two components: apprentices and mentors.

In the arts it is an accepted fact that a young person learns the trade best when studying with a master. In many segments of business, too, the importance of the mentor relationship for a young person's development has been documented. In systems development, however, this has not been a common practice. We suspect this is because (a) there are few masters available, and (b) it costs money. The success of an education program, however, is based on apprentices working with mentors.

By *mentor* we mean people who are *outstanding* in their field, not just good. They train by example, coaching their apprentices on a day-to-day basis to convey the knowledge of how to apply the tools the apprentices have already studied formally. The objective is twofold:

1. to deliver a good (system) product;
2. to transfer the skills of analysis (i.e., the analysis of processes and the analysis of data) and the skills of design.

Mentoring should not be a part-time effort. Mentors must never be brought in for intermittent consulting or to review or audit a product after the fact. They must put in a dedicated effort at the front end of a project. The mentor must be a legitimate member of the Development Team. For instance, the mentor would be responsible for the higher levels of analysis and the apprentice would be responsible for the lower, less complex levels.

Assuming an organization has more than one project going on at the same time, this level of involvement prevents any *one* person from being the mentor on every project during the initial training period. The mentor can't be everywhere at once. So, to advance on more than one front, import mentors. Imported mentors can be consultants who do the higher level work and transfer skills to the apprentices. It provides great training and usually delivers good products.

Even though you may choose a specific school of structured methodologies, your organization will probably evolve its own dialect within this school. As a result you will need someone to ensure that everyone in the organization using these techniques speaks the same dialect.

Similarly, you will need someone to ensure a similarity of approach among mentors who may be of somewhat different schools, and who may be reporting to different managers. This person's role in the mentor program is to:

1. help determine when a mentor is needed;
2. participate in the selection of mentors;
3. coordinate intragroup communication and tell them which dialect to speak.

Mentors can come from inside or outside the organization. Importing mentors is fine when you are in the process of grooming your own in-house. And each mentor does not need to be a master of all methodologies. Quite often, especially with imported mentors (i.e., consultants), they will specialize in either information modeling or process modeling or some other specific methodology.

Among the benefits you can expect from such a training program are:

- the job will be done better and faster;
- you will develop people better;
- the cross-fertilization of ideas will benefit everybody.

CHAPTER 7

MANAGER'S NOTEBOOK

7.1 DO MORE EFFECTIVE WALKTHROUGHS

Most companies have been walking through their delivered code for some time now. Naturally, we should all continue doing this. However, we have found that the most effective and productive walkthroughs are of analysis and design deliverables. That is, small, contained, and manageable walkthroughs of parts of the Information Model, Process Model, and Design Blueprint.

But these walkthroughs should always be with the affected or expert users; they should never be held in isolation somewhere in the corner of the data processing department. Walkthroughs are not presentations. And walkthroughs shouldn't be a case of meeting because the boss said so, and then filling out a checklist. That's not productive. Walkthroughs are peer quality assurance reviews by members of the Development Team, including users. The idea is to get the product right, to do good work. Real "let's-figure-this-out" working walkthrough sessions can ensure that the end product is about as right as it will ever get.

7.2 USE TEAMS

To get the best results possible, encourage the team approach to systems building.[1]

Never have just one person working on a development project! As hard as it may be to believe, there are countless data processing departments around the world who have several development teams, each consisting of *one* person. That person is business analyst, systems analyst, designer, data base specialist, communications expert, programmer, documentation specialist, controls expert, screen dialogues guru, and even test planner and tester.

No doubt, that single person—and there must be thousands of companies with such an expert—develops nothing but perfect systems! No doubt, this is why most systems deliver what the user wants, when she wants it, and at the right price.

Time and time again, it has been proven that the bottom line for productivity and quality is team capability. The major cost driver in Barry Boehm's COCOMO model,[2] exceeding all others, is team capability. Whether a company wishes to concentrate its productivity and quality improvement efforts on analysis, design, documentation, or coding—they all have the same source: the capabilities of the people doing the work.

Ability is developed by basic education and then by experience. We've repeated that fairly often, so you must be getting the idea it's quite an important issue with us. The slow way is to let people learn from their mistakes and to reinvent a few wheels along the way—the same mistakes that more experienced people have already made and learned from. The faster way is to mix the inexperienced people (with a firm footing in the appropriate theory) with the experienced ones, not just in the same room but on the same project. That is, create a Development Team of experienced people (with the most senior taking the role of mentor) as well as inexperienced staff (as students, perhaps at different levels of apprenticeship).

Team members naturally learn from each other. There is a synergistic effect, and the resulting deliverables usually reflect a quality that is greater than the sum of bodies would have suggested. Also, especially in analysis, the team approach helps combat the "blank page" syndrome, in which the lone analyst can't decide where to begin or gets bogged down or off track in the middle of analysis.

Great tools and insufficient skill levels will always have the same effect; there will be marginal, if any, improvement in quality. But the inverse is also

---

[1] One of the nicest little books written on the subject is Philip C. Semprevivo's *Teams in Information Systems Development* (New York: Yourdon Press, 1980).

[2] Barry W. Boehm, *Software Engineering Economics* (Englewood Cliffs, N.J.: Prentice-Hall, 1981).

true. Average techniques with great skill levels, and some imagination, will produce substantially better results. The greater payback results from investing in the people who must do the work, helping to develop their skills, expecting the best and settling for nothing less. And this can be done very effectively by developing a team approach to development projects.

Eventually, such highly trained, skilled, and motivated systems professionals will be much better paid than their industry peers. But because their performance level is so much better than average, fewer of them will be needed to do the same amount of work. Also, because quality will be so much better, there will be less repair maintenance as systems get older (and less front-end wheel spinning trying to figure out what the thing does). And because of staff experience, response time to change and enhancements will be significantly improved. All of which will free up more time and resources for new business development.

Undue optimism? No. A real goal that we believe can be achieved.

Develop teams, invest in their timely education, live by specific methodologies rather than none at all, and commit to excellence. And, with all of this, believe that you can achieve the best delivered products with the best people, and that this can start happening now.

Only your team of qualified, motivated people can deliver what the user wants, on time, and at the right price. Technology will not.

7.3 CONDUCT FORMAL POSTPROJECT REVIEWS

Do this with all Development Team members, including the users. (There's a tendency to be more honest and objective when users are involved.)

During the postproject review determine why things went well or did not. Discuss solutions for things that didn't go so well, so they can be done better next time. Make sure that the things that proved successful are done again. And spread the word about the good and the not-so-good. Compare actual time taken with the estimates. Only by measuring estimates against results will an estimator gain the expertise necessary to make better estimates in the future. Issue a memo and advise others of the actions that you took that improved or impeded productivity, quality, and the final level of success.

7.4 CREATE EFFECTIVE DOCUMENTATION

Although we must implement a new philosophy of how to go about doing the front-end work, the single most effective action we can take immediately is to document what we have so we can fully understand it.

But documentation must be effective, not conventional.

Conventional documentation tends to be wordy, bulky, and self-contradictory. It focuses on a low-level presentation of myriad details. Maintaining such documentation is a challenge on a par with maintaining the system itself. Even if it once included a high-level overview of the system, chances are that the overview, if it can be found, is five years out of date.

In one typical example, the original "physical models" of a system did not uncover all existing interfaces with other systems and departments, nor the effect of various actions on essential business events and responses. There was no data dictionary. Had work proceeded on this basis, the new system would have been incomplete, and much of the real work would end up being a complicated maintenance activity.

Because the existing documentation was inadequate, it was necessary to create new documentation of the existing physical system, without getting mired in details. After all, if the existing system is going to be replaced or changed substantially, we don't want to spend a lot of time and effort recording all its foibles for posterity. So the project team quickly created a set of high-level data flow diagrams, augmented by one additional level of more detailed diagrams. Informal supplementary notes were made for the narrative and data dictionary components of the "current physical model" (or *current* Functional System Model).

This high-level documentation was sufficient to ensure an understanding of the system, and became the starting point for developing an Essential Business Model. In a word, the result was lean and commercially usable.

Update Documentation Now

Ensure that the documentation is updated while software repairs or updates are in progress, not sometime in the future "when there is time." There is never time in the future.

There are three kinds of documentation intended for three different audiences:

1. The various models for systems developers and users (i.e., Essential Business Model, Functional System Model, Design Blueprint).
2. Operational instructions and procedures for DP Operations, DP security, audit, and so on (i.e., Operations Support Procedures).
3. Operational instructions and procedures for end users (i.e., User Documentation).

In the first category, these models are the system, and the only way to make intelligent changes is to start with the models, not to play catch-up after the fact. In other words, documentation is not a separate process, it is an integral part of the systems development effort.

Sec. 7.4　Create Effective Documentation　　　　　　　　　　　　　　　　**197**

In the other two categories, changes must be ready to install concurrently with installation of the new software. Therefore, these updates should be planned as soon as the nature of the changes is known (i.e., when the updates to the models are done) and prepared concurrently with coding and testing of the software.

Document Differently

Many people believe the Wharton Business School adage that "if isn't written down, it didn't happen." This kind of court reporter's practice might be followed for political reasons where the climate promotes a self-protective school of documentation. However, it is not appropriate for systems documentation and has diverted attention away from what really should be documented.

So we must stop the common practice of documenting, for the sake of history, everything that is encountered or happens. Keep only the business and functional models (with data dictionary and text), the code, procedural definitions for users, and necessary operations material. Only document and keep that which will be maintained and changed—nothing else.

Acquire a Documentation Specialist

Each systems development area should have a documentation specialist/technical writer resource who consults actively to other members of Development Teams. This person needs a working familiarity with the analysis and design tools being used, in order to:

1. be an active standards bearer, by participating in walkthroughs and informal reviews of project deliverables;
2. ensure communication is something that resembles English by reviewing, editing, and providing constructive feedback;
3. determine maximum and minimum documentation needs for any project;
4. establish a failsafe documentation upgrade procedure (for users and DP) for each delivered system;[3]
5. integrate documentation across related systems, by eliminating duplication and redundancies, resolving conflicts and ensuring consistency, being a liaison with data administration, and the like;
6. be a consulting resource re:
   - documentation needs and methods
   - interdependencies across related systems

The documentation specialist should never assume responsibility for writing the narrative and data dictionary components of systems models, because

---

[3] Possibly an audit every six months to ensure that the system's documentation still reflects what the system actually does—and a resulting action plan to get the documentation into step with the system.

this must, by definition, be done by the analysts and designers themselves. The documentation expert does, however, work closely with the systems creators to ensure that these textual deliverables are readable, understandable, concise, complete, unambiguous, and nonredundant.

Particularly in companies that develop and sell software commercially, the documentation specialist should work closely with the technical writers to ensure the highest possible quality of customer documentation. Professionalism is essential; in today's competitive market, it is no longer acceptable to foist a mess of incomprehensible gobbledygook on to the poor unsuspecting user.

It may be that in small companies the documentation specialist is the only technical writer around. Theoretically, there is no reason why a person cannot wear two hats, so long as it is clearly understood which hat to wear for which occasion.

7.5 PROJECT MANAGEMENT AND CONTROL

Throughout the text so far we have been focusing exclusively on those tasks that contribute to the completion of a project. In fact, the core of the superset approach is that of doing only the necessary work, and nothing else.

Real projects, however, are encumbered with issues and tasks relating to project management. On smaller projects, the person charged with project administration may also be doing a large part of the productive work, while on larger projects a significant number of people may be exclusively involved in running the project.

The systems development model that you abstract from your superset is a dynamic, iterative, three-dimensional network model of the processes you feel are necessary to complete your project. It is *not* a PERT chart or a list of checkpoints. If you try to draw a critical path through it, or put dates on the activities, you will have destroyed its flexibility and reduced its usefulness in adapting to future discoveries and changes in the course of the project. You may *version* or partition your system in various ways, taking various strands through different processes at different times, often repeating processes until you are satisfied.

The project manager must work with a different set of tools (such as PERT charts) and with a somewhat different objective. The Development Team is responsible for meeting user needs. The manager is responsible for having the user's needs met on time and within budget. These two objectives are quite different, and even if we must wear both hats on a given project, we must always understand which hat we are wearing at any point and time.

The SuperSet model assists the manager because it helps to break the project into smaller deliverables. With smaller tasks the manager is much sooner able to tell how things are going. In a classic development project with four enormous phases, the project may be dead in the water before the manager has a clue. If project tasks are decomposed to the level such that the manager sees tangible results from each person on the Development Team every couple of weeks (or less), then the manager is in a much better position to refine the expected completion date based on current estimates, to assign resources, and keep the user informed of current progress.

The productivity of the Development Team is ultimately a responsibility of project management, and here the SuperSet model can help again.

One of the dangers of a rigid life cycle model (or no model at all) is doing too much. That is, doing things that aren't needed at all (such as bureaucratic barnacles from previous projects), and overdoing some of the necessary activities. Adapting a development model for each project, and by dynamically refining it as the project continues, helps the project manager to maximize productivity by constantly focusing on essential work and ensuring that the right tools are used for each task.

Targets, Deadlines, and Estimates

Many projects have run into serious problems because project managers (or users) have confused targets, deadlines, and estimates.

Target: That date by which the user and project manager would like, and expect, to have the project implemented.

Deadline: That date by which it must be implemented.

Estimate: An assessment by a member or members of the development team as to when a task or series of tasks will be completed.

The target is derived by allocating a certain level of resources against the estimates. The target can be changed by changing the resources available (the project manager's job), but only those doing the work can change the estimates.

Users and managers cannot make estimates! They must be given willingly by the Development Team in response to open and unloaded questions. It is entirely a management responsibility to coordinate the estimates given by the Development Team with the user's demands. A systems developer cannot change how long it takes to do a given task! He or she can only do less of it, or do it poorly, neither of which are conducive to successful systems. A manager faced with estimates (either at the beginning of a project or revised estimates in midstream) that would extend beyond the target date has three alternatives:

1. add more resources (within limits);
2. change (or kill) the project;
3. move the target.

The SuperSet model doesn't deal specifically with the issue of allocation of resources because it is an asynchronous model. It is the manager's job to balance the requirements of this model with the real world of personnel, time, and budgets.

The accuracy of estimates may be in doubt, but the integrity cannot be. Estimating is a skill in its own right, but one that is learned fairly quickly as long as:

- the people making the estimates understand that their estimates must be honest and objective, and that they should not be intimidated by others;
- the task to be estimated is small;
- estimators are always given feedback on the accuracy of their estimates throughout the development process, and at postproject review time.

Metrics and Productivity

When we talk about productivity we must be careful of the context. There are many large efforts currently in place that will eventually help programmers do more work in less time. For example, fourth and fifth generation computing languages, artificial intelligence, and expert systems will help to deliver code perhaps hundreds of times faster than more traditional methods.

But when we are talking about programmer productivity gains of 100 or even 1,000 percent, it may not mean all that much in the overall development process. Recent projects have shown that coding represents between 10 and 20 percent of a completed project. The major effort still goes into figuring out the business events, the essential data, and the functionality of the system. While very desirable, huge productivity gains on the programming front still will not make the development of good, maintainable systems happen. Fast code is not a substitute for thinking—although there seems to be an industry-wide appeal for code prototyping as an excuse not to do analysis and design.

Since 80 percent or more of development effort is invested in analysis, design, and documentation, it would seem that these are the areas where the greatest and most significant improvements can be made.

To make sure we are all addressing productivity in the same context, let us define what we mean by productivity.

Productivity is a combination of lowering the unit cost of delivered products and maintaining or improving an acceptable quality level and delivery schedule.

Quality is flexibility, understandability, and the absence of defects.

Improved productivity only occurs when systems professionals provide better service to their clients, and when they are perceived to do this. The real measure of productivity entails meeting deadlines and providing quality systems in a more cost-effective way.

Since Tom DeMarco's public introduction of structure analysis in 1976, and some interesting enhancements since, there have been few practical efforts in how to do analysis and design faster and better (event analysis being the exception).

A lot has been said about the value of prototyping and fourth generation application generators. But the fact of the matter is that these excellent tools do not significantly speed up or improve the quality of necessary front-end work—they deal almost exclusively with transforming the desired functions into an automated implementation, usually on a single processor. The system outside the automated processors is commonly ignored. Needless to say, there is more to a system than its automated component.

It's our feeling that there have been few results in improving the speed and quality of front-end analysis, with the exception of event-based analysis—especially when there is a need to look beyond a simple application and at an entire way of doing business—because we are dealing with skills of synthesis, understanding, and imagination.

Although many members of the scientific community are trying, we have yet to develop the artificial intelligence to simulate these very human attributes. Yet, ever looking to technology to provide an answer, we say we *ought* to be able to do these things as fast as a computer can process data!

Unfortunately, although the human brain has more memory cells than the largest computer, it is a slow and imperfect processor when compared to the technology in use today.

Because the human processor is so slow and often imperfect, it is almost impossible to model. Most scientists accept that there is no immediate prospect of creating technological models of synthesis, imagination, and understanding.[4] With this acceptance (and a lot of marketing savvy) has come an incredible drive to improve productivity in areas we know our technology can do it: *the 20 percent or so of the systems development effort that relates directly to the machine.* So, while front-end business and systems analysis are almost completely ignored, almost all popular productivity effort is invested in those areas that are less esoteric and technologically easier to affect.

This is why almost every paper published today, every group discus-

---

[4] There are longer term prospects, however, with developments in the artificial intelligence research community, especially in the area of inference engines. Two particularly interesting articles were published in the August 1984 issue of *High Technology*: "Why Can't a Computer Be More Like a Brain?" by Eric J. Lerner, and "Computers and the Brain: An A. I. Perspective," by Johnathan B. Tucker.

sion held, every company debate started, is about code and its delivery—fourth generation languages, super high-level languages, smart application generators, programmer-less code, user-friendly prototyping, and so on. The technology that delivers the code is widely debated and heralded, but little is said about the cause of all improvements in technology: people.

The project manager should not be mesmerized by these debates and claims and expect such tools to miraculously deliver a system that represents what the user wants, on time, and within budget. The upfront work still has to be done.

Since it is generally accepted that the measurement of human productivity related to inventiveness, synthesis, business comprehension, and a multitude of other qualitative issues is no mean feat, almost all metrics today involve measuring or estimating the cost and delivery of lines of code (the notable and *meaningful* exception being *function point* analysis). However, we feel strongly that significant productivity gains will only be realized from overall improvements in quality. While quality may not reduce the cost of delivering code, it most certainly will enhance the lifetime maintainability and cost-effectiveness of the business unit. Our position is simple (and cost-effective)—if the user isn't happy, measurement has little meaning!

By not measuring productivity we will of course always beg the question of how we will know when we have improved it. We suggest you look to the effectiveness of your systems, how well they support the business, and to your users. Ask them how happy they are with your productivity. Are you more competitive and responsive to your firm's market? This may not be very scientific but it is an effective, inexpensive measure of whether or not you are *perceived* to be doing the job well.

The *real* measure of a valuable system is whether or not its cumulative cost is more or less than its accumulated lifetime benefits.

With *new* systems, then, we can certainly measure the pre-delivery *project* and postdelivery *product* costs, until system retirement. A formula that will produce a *quality indicator* (QI) by which we can measure different systems is:

$$QI = \frac{\text{accumulated lifetime benefits} - \text{accumulated defect costs}}{\text{total cost of building} + \text{enhancement costs}}$$

If the QI for a system is greater than 1, then the system was worth building. If the QI is less than 1, this implies a mistake in the development or a deterioration during enhancement. Either the benefits were overestimated or poor work has required excessive repair. This kind of measurement knows no automation boundary, is not tied to today's technology, and is based on the overall cost of supporting the business. It is also inexpensive (you already have all the data you need); and there is no danger that the measurement process will become more important than the job of delivering good business systems. It does force meaningful cost-benefit analysis, however.

There are some good books on measurement and estimating. We strongly suggest both Tom DeMarco's book on the subject[5] and Barry Boehm's,[6] who has many years of experience with metrics at TRW. The most recent and valuable addition, *Programming Productivity* by Capers Jones,[7] lists some twenty or thirty factors affecting productivity, as well as giving a very thorough discussion of the various metrics and their advantages and disadvantages.

7.6 PRODUCERS AND DIRECTORS

Do you remember the way credits roll by on the screen before every movie (the big names) and after (the little ones)? Something like this:

Produced by

Martin Megabucks

Directed by

Ted "Rambler" Yento

Starring

A. Nalyst
D. Ziner

and introducing

CHAD PROGRAMMER

as

"The Implementor"

Now, it isn't every day we turn to Hollywood for inspiration—after all, Q: What's the difference between Hollywood and yogurt? A: Yogurt is a live culture—but maybe they have something here.

[5] Tom DeMarco, *Controlling Software Projects: Management, Measurement & Estimation* (New York: Yourdon Press, 1982).

[6] Boehm, *Software Engineering Economics*.

[7] Capers Jones, *Programming Productivity* (Englewood Cliffs, N.J.: Prentice-Hall, 1986).

Often in the business world we allow ourselves to be restricted by the hierarchical nature of our corporate cultures. All the people associated with a given project can usually be seen as part of an hierarchical structure with the project manager at the top and the people doing the real work near the bottom. In fact, this narrow view has led to disastrous mistakes in the past, such as promoting good programmers to the position of analyst, or promoting good analysts to the position of project manager.[8]

In the theatrical environment there are two vitally important senior roles: producer and director. Of these, the producer's role is most similar to the current responsibilities of a typical project manager—controlling the resources and ensuring the delivery of a product on time. The director is responsible for making sure the product is of high quality.

A similar approach works very well with larger systems development projects. The project manager deals directly with the clients and is responsible for personnel and project administration. The director, or *project architect*, is responsible for making sure that the right tasks get done, using the right methodologies and tools, and for providing support and guidance where needed.

This is not another new level of bureaucracy to be added to a project. That won't get the job done any better or faster. However, as on a construction project or on a movie set, the chief architect or director stays with the project from the beginning to the very end, bringing all the loose ends together and molding the product into his vision of a quality deliverable. The architect could also very well be the mentor we discussed earlier. The producer's job (or project manager) is to ensure that the product is profitable. This is not always an easy feat, since architects and directors sometimes feel that quality is not inexpensive!

Therefore, the project manager and project architect together must ultimately be responsible for delivering a cost-effective, profitable system to the user. The distinctiveness of the two roles may force into the open a debate between virtue and expediency that would otherwise be suppressed (or ignored).

It's interesting how the idea of chief architect isn't really all that new. Many years ago the concept of the chief programmer was introduced. It's hard to tell how much of the development cycle was devoted to programming way back then, but today it's somewhere between 10 and 20 percent of the entire effort. The balance is analysis and design. It makes sense, then, to assign a chief architect to critical projects (aren't they all?) from the beginning to the end!

---

[8] "*Degrade* v.trans. To promote (a programmer) to systems analyst." From *The Devil's DP Dictionary*, by Stan Kelly-Bootle (New York: McGraw-Hill, 1981).

7.7 AUTOMATED TOOLS FOR SYSTEMS DEVELOPMENT

The biggest single missing item in this book so far is a discussion of automated tools for systems development.

Software developers—the very people who provide the most technologically advanced solutions to other people's problems—have always had to rely on low technology (pencil, paper, and eraser) to do their own work. While the developers of technology have introduced almost everyone else to the term "computer assisted," the very industry responsible for this important innovation has, paradoxically, continued to design and build software the old-fashioned way—slowly, laboriously, by hand. But now, after a long wait, systems developers can use technology to help do what we have always done for others—make life easier.

It simply made no sense to continue analyzing and designing systems with a pencil and eraser at a desk that had a computer on it! So a multitude of software firms have finally entered the market with various versions of automated tools for systems development. Stand-alone tools, such as the Analyst/Designer Toolkit (Yourdon, Inc.), ProKit*ANALYST (McDonnell Douglas), AUTO-ASYST (Atkinson Tremblay Assoc.), and many more, are now competing for market share from professional systems developers.

These tools will have a dramatic impact on the systems analyst and designers and programmers in the near future. (The objective of all of these tools is to be able to ultimately deliver *compilable* specifications.) Right now, they all make it a lot easier to produce and maintain the necessary models and associated systems documentation. Most of these software products will provide some or all of the following features:

- reduce the *time* required to make changes to diagrams;
- *integrate* diagrams and the data dictionary, thus helping to ensure integrity across all systems documentation;
- collect data dictionary entries *directly* from the diagram;
- provide *automatic validation* to avoid duplicate names or entries in documentation;
- allow *interactive* change of systems information during analysis and design;
- check for *imbalances* across diagrams;
- *verify* that all data names are actually entered in the data dictionary;
- apply a change to a data name and the like *globally* to all occurrences of that name everywhere in the documentation;
- *reduce the effort* of documenting the result of analysis and design considerably;
- *reduce the effort of maintaining documentation* substantially.

By the time this book is widely circulated, most of the larger companies will be using integrated productivity tools that combine CAD/CAM techniques with project management methodologies—for example, customized versions of the Frantzen/McEvoy Superset model will be directly implemented on a computer system, and the activities of the systems analysts and programmers will be monitored and/or regulated by a mechanized project management system.

This is a rapidly expanding field that should be high on the priority list of any project manager. Find out what's available, who offers what, what kind of productivity gains you can have with such tools (some industry sources are claiming improvements of a low of 20 percent to a high of 900 percent!), and what you should be doing about it in your shop. Some of these tools can *help* deliver what the user wants, on time, and at the right price!

CHAPTER 8

SUMMARY

In summarizing, we focus on the most important aspects of the superset approach:

- If it ain't broken, don't fix it!
- Obey the "Lump Law"!
- Good people do good work.
- Adapt, don't adopt!

If It Ain't Broken, Don't Fix It

As systems builders, we often find ourselves believing that a new system will always be better than an old one; that change will always be for the better. Whether this is due to rampant egoism in all data processing professionals, or just a subconscious desire to ensure job security, the results often leave the user worse off than before.

We must learn to leave well enough alone.

The cost of doing nothing must always be honestly considered at the same time that other alternatives are being costed. The system not built is often the cheapest!

As we do with the development of systems themselves, we must focus on the process of systems creation and ask ourselves, "What activities are essential to the delivery of this system?" Activities that are not essential should be abandoned.

We must never do anything without fully understanding why we are doing it, and how the results of a given activity are needed by the user, or will move the project closer to completion. If asked, "Why are you doing that?" answers such as:

1. "Because we always do it"; or
2. "Because it's development life cycle activity 37"; or
3. "Because Frantzen and McEvoy show it as a process in their model of a systems development system";

are just not good enough. The answer must always begin with "Because it's needed by. . . ."

After-the-fact systems documentation (not user documentation, which is needed by the user) is probably one of the better examples of an activity that can be jettisoned. The specification documents and graphic tools used in the systems creation process are all the documentation you'll ever need.

Obey the "Lump Law"

Jerry Weinberg coined what has become known as the Lump Law (Fig. 98) "In order to understand anything, we shouldn't try to understand

Figure 98: The Lump Law

everything." Any system we want to understand, whether it is a business system or a systems development system, is too complex to be understood all at once. Only by partitioning the system into sufficiently small mini-systems can its components be understood—and by synthesis, the entire process.

One of the questions most frequently asked when we are teaching systems modeling is "How detailed should I get? How deep should the levels of my DFD go?" The standard answer in building a process model is "Until the specification can fit on one page, and can be easily understood."

Although we don't write formal process specifications for the SuperSet System Development Model, a similar guideline applies: it must be possible to produce a tangible deliverable from the process in a reasonable period of time, and the process must be well understood.

A slightly facetious but very useful rule of thumb would be that a systems development model that you create to apply to specific projects should be partitioned in sufficient detail so that no activity takes longer than the annual vacation allowed to the person doing the work. For newer people two weeks should be the maximum. More experienced hands might have some activities that take three or four weeks. But that's the maximum. Activities that only take a day or two are also fine.

Another way of looking at it is that no activity should take so long that a total disaster in doing the work for that one activity would, by itself, have a major impact on the project. Clearly, if an activity takes two to three times as long as expected, or if major difficulties are encountered, it's more than likely that other problems will surface. But at least the project manager gets fast and early notice that there may be trouble ahead; and that allows her to start considering changes that should be made.

It's also important that the chunks of work be small enough so that the people doing the work can forget about how long it takes to do it. That's a management issue, and only misleads the people doing real work. The focus must be on zero defect work—doing it right!—and not on getting it done by Thursday. Only if the pieces are quite small can the people doing the work be relieved of the management burden. If it takes till Friday to do what was estimated as a three-day task, and was expected on Wednesday, that's something for the project leader and the user to discuss.

Good People Do Good Work

The quality of work done by a Development Team is directly related to the training and experience of team members and to the quality of the tools with which they work.

The New Chrysler Corporation would never consider hiring an engineer fresh out of college and assigning her the responsibility for designing the new 1995 Scylla, even after a three-month in-house course on "How to Design and Build Cars" (although the old one might have!). Nor will

Chrysler try to build such a car using the tools and techniques it used twenty years ago. Any business must realize that its systems staff is a resource that must be developed and maintained.

Labor is the largest cost driver of our new systems, and the efficiency and quality of that component require a commitment to ongoing investment in development and support. The use of the SuperSet model alone won't help if the people using it aren't trained and experienced in the underlying tools and techniques.

The use of walkthroughs is one of the most important aspects of building expertise in your staff. People learn by seeing how real problems are solved. Moreover, though walkthroughs cost money, there is a double payback since they are the most effective mechanism for ensuring the delivery of quality products.

Adapt, Don't Adopt

This book that you are holding in your hands is a very concrete object. Its prose is narrative, linear, and procedural. What's unfortunate is that these qualities are the worst ones to bring to the creation of a systems development model for your project.

This is not a generic development model, because a generic model is almost useless and can cause as many problems as it solves. Rather, it is an example of a possible development model. It is essential to remember that the systems development process is iterative, not linear. Each activity must be reworked and refined until all parties involved are satisfied that the product of the work is of the highest quality.

In writing this book, however, we had to produce this sample model. Study it, make sure you understand what we mean by it—and then forget it! In other words: do as we say, not as we do!

The process of systems building—and of building systems building models—is fluid, dynamic, iterative, and recursive. Use the guidelines in this book, then adapt the models in light of your environment and circumstances.

If you are working from an existing methodology model, cheat! Do whatever possible not to deliver stuff. Always ask for each development activity, "Can the system still be delivered, fully functional, if we don't do this?"

You must build your own SuperSet model. Neither ours nor anyone else's will fit the specific needs of your business and environment. Also, once you have built a SuperSet model, it will have to be maintained and enhanced as the business and technology change around it.

More than ten years ago, Stafford Beer said: "With today's technology, we can implement anything that we can specify. The problem, then, is learning how to specify."

CHAPTER 9

THE DICTIONARY

> The Dictionary is in alphabetical order.
> All entries in UPPERCASE have additional detail and are defined elsewhere in the Dictionary.

ACCEPTANCE CRITERIA

What Is It?

The Acceptance Criteria of the target system complement the functions specified in the model of the system.

The Acceptance Criteria can be documented separately, much as a checklist, to specify exactly what the system will and will not do. The Acceptance Criteria checklist, if produced as a separate deliverable, must reflect all of the policies or rules documented in the INFORMATION MODEL and how these rules are executed as reflected in the PROCESS MODEL.

The Acceptance Criteria checklist should also specify how the system will behave in its physical environment. And, finally, this deliverable should specify how it will respond to any external physical constraint. For example, a payroll system producing tax withholding statements must produce them in the format dictated by the government.

Who Can Help?

- The DEVELOPMENT TEAM
- The QUALITY MANAGEMENT TEAM
- The OPERATIONS group
- Internal Audit
- Systems Security
- CLIENT staff

Who Might Need This?

- The QUALITY MANAGEMENT TEAM
- The TESTING TEAM
- Internal Audit
- Systems Security
- CLIENT management

What's Inside?

- Acceptance Criteria for all system functions, inside and outside the automation boundary
- Acceptance Criteria dictated by external sources
- ENVIRONMENTAL ACCEPTANCE CRITERIA

APPROVAL GROUP LIST

What Is It?

A list of all the people and the departments they belong to who can approve major milestones through the development life cycle—that is, approvals of ESSENTIAL BUSINESS MODEL, VERSION PLAN, the "automation boundary," the FUNCTIONAL SYSTEM MODEL, the TEST PLAN, and the SYSTEM RELEASE. These approvers are also often direct "users."

Chap. 9 Audit & Security Guidelines   213

Who Can Help?

- The DEVELOPMENT TEAM
- CLIENT management

Who Might Need This?

- The DEVELOPMENT TEAM

What's Inside?

- names
- titles
- areas of responsibility
- locations
- phone numbers

AUDIT & SECURITY GUIDELINES

What Is It?

Almost every company has some guidelines on security and audit to be considered during the development process. These guidelines should identify to the DEVELOPMENT TEAM general audit, security, processing, recovery, and environmental issues that are of concern to audit and security personnel.

Who Can Help?

- Internal Audit
- Systems Security

Things to Consider:

- Initial data capture and *identification* must be completed
- Control should be established close to transaction source and carried through the entire system
- Input must bear *evidence* of authorization
- Error prevention and detection procedures should be in place at the *input* preparation or processing stage
- *Completeness* and *accuracy* of data input must be ensured
- Procedures for early detection of hardware or software *malfunction* must be in place

- There must be methods to *measure* successful processing
- There must be appropriate use of edits
- There must be appropriate checking, distribution, and *reconciliation* procedures for data output
- There must be appropriate error *correction* and reentry procedures
- There should be procedures for *forward and backward tracing* and identification of data elements
- There should be separation of *incompatible* duties in the system
- There should be *periodic reconciliation* of subledgers
- There should be *periodic reconciliation* of the update activity
- There should be Suspense Account analysis and reconciliation
- Online system access must be restricted to those *authorized*
- Completeness and accuracy of data transmission must be *verified*
- There should be *online recovery controls* to ensure that data are not lost or duplicated
- There should be *receiving system controls* to ensure that data feeds are complete and accurate
- There should be appropriate *physical security* for
    physical site
    teleprocessing access
    hardware device protection (lines, modems)
    information distribution
    data store access
    passwords
    systems documentation
- There should be appropriate *software security* to control
    program or module access
    transaction security
    attempts at security breaches
- There should be sufficient security on *system maintenance* to detect and prevent system subversion[1,2,3]

[1] Allan Brill's *Building Controls into Structured Systems* (New York: Yourdon Press, 1983) deals with the controls issue perhaps more thoroughly than any other recent book and discusses how controls are illustrated in process models.

[2] J. Palmer and S. McMenamin's *Essential Systems Analysis* (New York: Yourdon Press, 1984) differentiates between the "model of essence," which is free of implementation methods, and the "implementation" model, which contains all edits, audits, time, and technology considerations.

[3] In the SuperSet System Development Model, audit and security considerations are introduced for the first time with the FUNCTIONAL SYSTEM MODEL. Many of the "rules" of control are detailed in the INFORMATION MODEL and reflected in the PROCESS MODEL. Additional design-related controls are detailed when preparing the DESIGN BLUE-

CLIENT (THE)

What Is It?

The person, or department, or group of people and departments that sponsor and pay for the contracted service or system to be developed. The CLIENT is more commonly called "the user."

CLIENT LIST

What Is It?

A list of all the people who are involved in the development project and have a direct or indirect reporting relationship with the CLIENT. These are the people more commonly called "users" or "user management."

This list usually consists of management level or key employees, since it generally isn't necessary to list all the clerical personnel the DEVELOPMENT TEAM will be interviewing during the development process.

These are the main people who will be involved in developing and reviewing the accuracy of project deliverables such as the COST-BENEFIT STATEMENT and INFORMATION MODEL, reviewing the TEST PLAN, and determining the "automation boundary."

The APPROVAL GROUP is usually derived from this list.

Who Can Help?

- The DEVELOPMENT TEAM
- CLIENT management

Who Might Need This?

- The DEVELOPMENT TEAM

What's Inside?

- names
- titles
- areas of responsibility
- locations
- phone numbers

PRINT. All of these control issues become ACCEPTANCE CRITERIA and are used to develop the TEST PLAN.

CLIENT TRAINING PACKAGE

What Is It?

A "package" used to train the CLIENT's staff in the support of the new system. This kind of training program is generally only produced with the more complex and sophisticated systems. In other cases "user training" is done during the development process, when user staff is part of the DEVELOPMENT TEAM and learns what the new system will do, including the processes outside the automation boundary.
This training package is incremental, as new versions of the system are developed and released.

Who Can Help?

- CLIENT staff
- Education department staff
- QUALITY MANAGEMENT GROUP staff

Who Might Need This?

- The CLIENT

What's Inside?

- general information:
    what manuals are available and where
    how to find things in the manuals
    how to ask for system changes
    who to call in case of difficulties
    how to report a problem
- hardware available to client staff:
    basic information on how to operate
    what to do and who to call if hardware fails
    how and when to reorder supplies
- dialogues:
    how to get data
    how to manipulate data
    how to manipulate screen formats
    what security features there are
- backup and recovery:
    what happens, when
    how long it takes to happen
    how data integrity is maintained

- functions:
    how to prepare data for the system
    how to get data into the automated system
    what to do with data coming out of the automated system
- error handling:
    what to do with calculation errors
    what to do with missing data
    what to do with just plain wrong data
    what to do when the automated system just won't go (for online systems)
    who to call when nothing else works[4,5]

CODED PROGRAMS

What Is It?

The implementation of the design inside the automation boundary of the target system.
For our purposes, this includes source code, executable code, and any listings associated with code.

Who Can Help?

- QUALITY MANAGEMENT TEAM staff

Who Might Need This?

- The CLIENT

CONVERSION SCOPE

What Is It?

A statement of what data and how much of the data are to be converted from the old database, files, records, or index cards to the new CONVERTED DATABASE. It should also specify what will *not* be converted by the DEVELOPMENT TEAM.
The Conversion Scope should also include priorities, schedules, deadlines, and budget allocations.

---

[4] A documentation specialist from the QUALITY MANAGEMENT GROUP could certainly be of considerable help with this.

[5] Although the client has the real need for a training package, in many companies the education itself is administered by an education department. It's also possible that the education department developed the training program with the help of the DEVELOPMENT TEAM.

Who Can Help?
- The DEVELOPMENT TEAM
- The CLIENT

Who Might Need This?
- CLIENT management
- The DEVELOPMENT TEAM
- Internal Audit
- Systems Security
- Data Entry management

What's Inside?
- a definition of what is to be converted
- conversion priorities (relates to VERSION PLAN)
- constraints, in terms of
    how much of specific data groups to be converted
    what data groups *not* to convert
    financial
    scheduling and target dates
    data privacy
- a definition of data criticality, usually resulting from risk analysis provided by internal audit and systems security staff

CONVERTED DATABASE

What Is It?

All the data specified in CONVERSION SCOPE, converted and resident on a new set of data stores. The new data stores and the resident data conform to the DATABASE DESIGN and meet all the goals specified in (Conversion) ACCEPTANCE CRITERIA.

What's Inside?
- all converted data as required by the target system

COST-BENEFIT STATEMENT

What Is It?

Initially, a report comparing the costs and anticipated benefits of four ways of approaching the ESSENTIAL BUSINESS MODEL. The four alternatives evaluated are: (1) Maintain the Status Quo (i.e., "do nothing"); (2) Do the Immediate Opportunities for Improvement; (3) Design a New System; (4) Buy Packaged Software. Evaluation of a software package is usually not appropriate immediately after the ESSENTIAL BUSINESS MODEL has been produced. But it is appropriate to evaluate software packages after the FUNCTIONAL SYSTEM MODEL has been delivered.

The idea is to select the alternative offering the best investment opportunity over several years to the client. The client is then advised of changes to the cost-benefit figures throughout the development life cycle, as costs become firmer and more is learned about the impact of the system on expenses and returns. The statement should normally cover at least five years.

This statement is refined as needed by the client during the development cycle. A final version of it is produced after implementation, based on actual development costs.

Who Can Help?

- The DEVELOPMENT TEAM
- CLIENT management
- QUALITY MANAGEMENT TEAM management

Who Might Need This?

- CLIENT management
- Project management

What's Inside?

- for each alternative—(1) maintain the status quo ("do nothing"); (2) do the IOIs; (3) design a new system; (4) buy packaged software—the following information should be provided:
- Management Summary page:
    (a) Operating Cost of Current System
    (b) Estimated Operating Cost for Version $n.n$
    (c) Benefits (a − b)
    (d) Discount Rate (at specified percentage)
    (e) Present Value of Future Benefits (c ∗ d)
    (f) Development Costs
    (g) Net Present Value of Investment (e − f)
    (h) Profitability Index
    (i) Return on Investment (as percentage)
- General Assumptions
- Analysis of Results of each alternative

- Detailed Cost-Benefit Analysis:
    - *Development* (from creating the ESSENTIAL BUSINESS MODEL through implementing the SYSTEM IN PRODUCTION; this includes cost for conversion of data)
    - *Hardware* (includes such items as terminal charges and personal computers)
    - *Data Processing* (includes production costs, data entry, maintenance of the new system and maintenance of the old system while migrating to implementation of the "final" version of the new system, and such things as TSO charges)
    - *Administration* (includes personnel, materials, and education costs)

DATABASE DESIGN

What Is It?

A complete description of all data elements and data structures in the system and a description of how the data structures will be implemented.

Who Can Help?

- QUALITY MANAGEMENT TEAM staff
- OPERATIONS group
- Database administrator

Who Might Need This?

- The QUALITY MANAGEMENT TEAM
- The DEVELOPMENT TEAM
- The TESTING TEAM

What's Inside?

- physical descriptions of all data elements and structures
- descriptions of any relationships between structures
- descriptions of the access method for automated data
- descriptions of the storage medium for all data

DATA DICTIONARY

What Is It?

A Data Dictionary is an ordered set of definitions of "things" found in or behind data flow diagrams[1] and entity-relationship diagrams.

The kinds of "things" found and defined in the Data Dictionary are:
- objects or entities
- the relationships between objects[2]
- data stores
- data flows
- data elements

Who Can Help?

- CLIENT staff
- QUALITY MANAGEMENT TEAM staff
- The DEVELOPMENT TEAM

Who Might Need This?

- The DEVELOPMENT TEAM
- The QUALITY MANAGEMENT TEAM
- CLIENT staff

What's Inside?

- data elements should be given precise, meaningful names that uniquely identify the unit fact being represented
- the meaning of data elements must be defined through explicit statements of business policy. These statements should cover:
    - role/purpose/function
    - refinement of the name
    - source/verification authority
- the data content or value ranges permissible for each of the data element components must be specified
- known data dependencies (on other data elements) must be specified:
    - composition
    - computational derivability
    - limits
    - existence

---

[1] Tom DeMarco discusses the data dictionary extensively, as it relates to data flow diagrams in chapters 12, 13, and 14 (pages 129–165) of *Structured Analysis and System Specifications* (Englewood Cliffs; N.J.; Prentice-Hall, 1979).

[2] Not everyone agrees with putting relationship definitions in the data dictionary, since the definitions are more textual and the format is therefore different. We have found it very useful, though, since the rules still have to be looked up somewhere and this eliminates the need for a second dictionary. It's our feeling that there should only be one data dictionary, rather than one for each kind of model. Having only one may be duller, but it makes life somewhat more comfortable.

DESIGN BLUEPRINT

What Is It?

The Design Blueprint is the "packaging" of all the information needed by programmers to build the automated portion of the system.

Who Can Help?

- QUALITY MANAGEMENT TEAM members

Who Might Need This?

- The QUALITY MANAGEMENT TEAM
- The TESTING TEAM

What's Inside?

- STRUCTURE CHARTS
- DATA DICTIONARY
- MODULE SPECS
- DIALOGUE DESIGN
- STATE TRANSITION DIAGRAMS[1,2]

DEVELOPMENT TEAM (THE)

What Is It?

The team of people associated with the analysis, design, and quality engineering of a client's system. It should include "user analysts." This team does not necessarily feed and care for the system after implementation, although it may in some shops.

Team members are not necessarily all full-time members of the development project. There should always be at least two full-time dedicated team members; others can assist as their particular skills are required.

Also, teams can often be different from activity to activity (or from one development phase to another). That is, analysis teams (which are heavily user-based) are staffed differently than design teams (which are oriented more toward technical skills).

---

[1] *Structured Design* by Edward Yourdon and Larry L. Constantine (New York: Yourdon Press, 1978) is a classic in software engineering and should be read by everyone.

[2] *The Practical Guide to Structured Systems Design* by Meilir Page-Jones (New York: Yourdon Press, 1980) is the first book to link the concept of structured design with the techniques of structured analysis. Designers working with process models will find Page-Jones's work invaluable.

An example of a design team could be:

Project Team
Giselle Werbezerk-Piffl	(Project Manager)
Kenneth McEvoy	(Chief Design Analyst)
Horacine Clutch	(Design Apprentice)
Ovid Futch	(Design Apprentice)

Internal Consultants
Trond Frantzen	(Quality Management)
Verbal Funderbuck	(Documentation Consultant)
Humperdink Fangboner	(Controls/Audit)
Siddhartha Rosenbloom	(Database Consultant)
Thusnelda Neusbickle	(System Security)
Rommie Vanderboor	(Strategic Planning)

The *"Project Team"* are the full-time members, while the *"Internal Consultants"* provide expertise as needed. Also, individuals are listed by function performed rather than by title and reporting relationships. Ignoring titles and concentrating on functions usually enhance team cohesion.[1]

The "team" concept is based on the fact that any one or two individuals cannot possibly know everything there is to know about developing a system. Also, it takes advantage of the synergistic effect a team can provide.[2]

Attributes:

- ability to be *analytical*; an appetite for tearing systems apart plus a merciless desire to scrutinize each piece;
- ability to resist doing something because it is "convenient";
- commitment and ability to communicate clearly with other humans, not just machines;
- ability to tolerate uncertainty, to take satisfaction in identifying *questions* as well as recording answers;
- ability to withstand the abuse that always accompanies any effort to "do good work";

[1] In *People and Project Management* (New York: Yourdon Press, 1980), Rob Thomsett describes how to develop a successful project management system that facilitates the flow of information among analysts, designers, implementers, maintainers, and users while taking into account the needs of these individuals. His book discusses individual motivation, job satisfaction, performance, work attitudes, leadership roles, management issues, and how new software design methods have affected control of systems.

[2] A methodology for applying strategies based on management theory and organizational development theory to the special problems of systems development groups is described by Philip C. Semprevivo in *Teams in Information Systems Development* (New York: Yourdon Press, 1980).

- ability to distinguish fundamental business policy from issues of implementation and then, during design, implement the business policy in the most flexible, professional manner possible;
- ability to use the tools of analysis and design instead of being a slave to them; to recognize their limitations as well as their strong points;
- willingness to *invent* whenever the need arises.

DIALOGUE DESIGN

What Is It?

The Dialogue Design is a model of the screen I/O side of an online system. It will describe all the screens in the system and the relationship between the screens, and it will identify the application functions that will be needed to support the screens.

Who Can Help?
- QUALITY MANAGEMENT TEAM staff
- CLIENT staff
- Shakespeare
- Technical experts in dialogue design

Who Might Need This?
- The QUALITY MANAGEMENT TEAM
- The TESTING TEAM
- The DEVELOPMENT TEAM
- CLIENT staff

What's Inside?

a model of an online session, showing all screen layouts. For each screen, there must be:
- a description of any predisplay or postdisplay functions needed to support the screens (user exits)
- a description of what user response controls the selection of the next screen, if there is more than one option; these can also be expressed with STATE TRANSITION DIAGRAMS for some systems.

If a scenario modeling tool is used, the scenario is sufficient description of the screens and the control of the dialogue flow, but the screen chart and the list of support functions must still be produced.

DISASTER RECOVERY PLAN

What Is It?

This is a document describing what to do when a disaster strikes.[1] What this *really* means is: What do you do to restart operations after the loss of the entire data center facility? The DEVELOPMENT TEAM, with a few others, should work out the answer to this question and document it here. Absence of a workable, documented plan could be debilitating to the company. Most of this work will likely be done by the OPERATIONS group.

Who Can Help?

- The DEVELOPMENT TEAM
- The QUALITY MANAGEMENT TEAM
- CLIENT management and staff
- OPERATIONS group
- Internal Audit
- Systems Security

Who Might Need This?

- Systems Security
- OPERATIONS group
- CLIENT management
- QUALITY MANAGEMENT TEAM management

What's Inside?

- Where is secured systems recovery documentation?
- Who is in command?
- Where are secured backup databases, files, and software?
- Where are the alternate data centers?
- What are the priorities?
- Who is responsible to do what?
- How do you reach those people and what do you tell them?

---

[1] If disaster strikes outside the automation boundary, that is, at the nonautomated part of the system, it's generally accepted that the client should have her own set of Disaster Recovery Plans to deal with the problem. It is, however, the DEVELOPMENT TEAM's responsibility to advise client management of the need for this documentation covering nonautomated parts of the system.

- What government agencies need to be notified?
- What do the government agencies need to be told?

DOCUMENT REQUIREMENTS

What Is It?

Defines all the physical requirements for any special forms, reports, or documents needed by the target system. It's intended that this deliverable should be a specification rather than a design. Design of documents and forms can be "contracted out" by the DEVELOPMENT TEAM to an Office Systems group or a Forms Design group. Systems development people have never been known to be too good at this kind of stuff anyway!

Who Can Help?

- The DEVELOPMENT TEAM
- CLIENT staff
- Forms Design group
- Office Systems group
- Systems Security

Who Might Need This?

- Forms Design group
- Office Systems group
- Systems Security
- Internal Audit
- CLIENT staff
- The QUALITY MANAGEMENT TEAM

What's Inside?

- size requirements (dimensions)
- preferred paper quality
- type of hardware to be used with the document
- color needs
- number of copies
- source and distribution of the document

- sequencing or numbering, or other audit and security needs
- other language requirements
- legal requirements
- retention method
- special needs for filing
- special needs for long-term storage
- volume estimates for a given period
- peak period projections
- projected increase or decrease of use over a given period
- retention period
- special compatibility needs of the outside world (postal, banking, and securities requirements)
- sample (prototype) of the document

DOCUMENTATION HEAVEN

What Is It?

A shining and ethereal place, reputed to be at the end of the rainbow, spoken about in hushed and reverent tones by data processing serfs and churls around the world.[1]

It is said that Documentation Heaven will hold incarnations of all of life's systems—incarnations old and new, good and bad, useful and not. But it is also said we mustn't let the vast number of incarnations fool us, for beyond the multitudes of physical transformations will be found True Essence. It is simply a question of looking within.

The Keeper of Good Things will surround STUFF KEPT AND MAINTAINED with Goodness and Light for all time, if it is Maintained. However, some alleged visitors to Documentation Heaven have returned with warnings for us: unless THE STUFF is maintained by the Original Creators or those who come after Them, it will pass out to an Unreachable Nether World. There it will serve as a False Beacon, deceiving people into believing that it is still Good Stuff.

Paradoxically, The Keeper will allow some STUFF to be Kept in Heaven that was never intended to be Maintained, but has Inherent Goodness. That is because Project Managers and Other Bureaucrats have access to Documentation Heaven too. (The things some Keepers will do to please Project Managers.)

Documentation Heaven is a Good Place.

---

[1] We are not referring to the Senate.

What's Inside?

- STUFF KEPT AND MAINTAINED

ENVIRONMENT TESTING PACKAGE

What Is It?

This is the software and procedures, and their documentation, for testing the system's response to various environmental conditions. It is primarily based on the ENVIRONMENTAL ACCEPTANCE CRITERIA.

This testing package must define the testing procedures and expected results for both automated and manual procedures.

An environmental test may consist of procedures to determine that the system behaves in a reasonable fashion when encountering the following problems:

Volume: Heavy loads over a long time span
Load/Stress: Heavy loads in a short time span
Hardware: The physical components of the network
Security: Attempts to break the system's security
Performance: Required response time and throughput rates
Usability: Human impact such as comprehension of diagnostic messages and screens/reports (i.e., "user friendliness")
Reliability: Minimum availability of system
Recovery: (1) Recovery from hardware failure and length of time to recover; (2) recovery from a disaster and how long it takes

Who Can Help?

- The QUALITY MANAGEMENT TEAM
- CLIENT staff
- The TESTING TEAM
- Internal Audit
- Systems Security
- OPERATIONS group

Who Might Need This?

- The TESTING TEAM

What's Inside?

- test package procedures and software, including identification (e.g., a policy or security number)

why it is included (with specific reference to the ACCEPTANCE CRITERIA)
expected results
things that have to be done (i.e., a tasklist)

ENVIRONMENTAL ACCEPTANCE CRITERIA

What Is It?

A definition of how the system must behave under specific environmental conditions.

Who Can Help?

- QUALITY MANAGEMENT TEAM
- The DEVELOPMENT TEAM
- OPERATIONS group
- Internal Audit
- Systems Security
- CLIENT staff

Who Might Need This?

- QUALITY MANAGEMENT TEAM
- The TESTING TEAM
- Internal Audit
- Systems Security
- CLIENT MANAGEMENT

What's Inside?

specific goals (i.e., acceptance criteria) for testing purposes, covering the following:

- *Volume*: Heavy loads over a long time span
- *Load/Stress*: Heavy loads in a short time span
- *Hardware*: The physical components of the network
- *Security*: Attempts to break the system's security
- *Performance*: Required response time and throughput rates
- *Usability*: Human impact such as comprehension of diagnostic messages and screens/reports (i.e., "user friendliness")
- *Reliability*: (1) How to recover from hardware failure and length of time

to recover; (2) how to recover from a disaster and how long to recover.[1]

ESSENTIAL BUSINESS MODEL

What Is It?

The Essential Business Model is simply the "packaging" of the components of the ESSENTIAL INFORMATION MODEL and the ESSENTIAL PROCESS MODEL. The data dictionaries for each of these models are merged into one DATA DICTIONARY.

Who Can Help?

- The DEVELOPMENT TEAM
- The CLIENT's senior staff
- The Data Dictionary Administrator[2]
- The department's secretary

Who Might Need This?

- The DEVELOPMENT TEAM
- The CLIENT
- Corporate Systems Planning

What's Inside?

- (refined) OBJECTIVES & CONSTRAINTS
- ESSENTIAL INFORMATION MODEL
- ESSENTIAL PROCESS MODEL
- DATA DICTIONARY (combined, for both preceding models)

ESSENTIAL INFORMATION MODEL

What Is It?

The essence of a system consists of all aspects of the system that *must exist* regardless of the level of technology used to implement the system. The essence of a system consists of:

[1] See also DISASTER RECOVERY PLAN.

[2] Documentation will soon get completely out of hand if everyone in the company, or even just the department, can use any data names, object names, and definitions he or she pleases. While not quite a Database Administrator, a data dictionary administrator can be an invaluable asset to "standards" development, and certainly an investment in the future (the quality issue and all that). You should consider developing one.

- its essential *activities* and
- its essential *memory*.

It is a fundamental premise that before a business model can have any meaning, the rules, conventions, and laws that govern the operation of the business must be understood and defined. To this end the Information Model is descriptive, rather than procedural; it describes business subject matter in terms of the business *entities* and *relationships* that make up the subject matter.

The Essential Information Model[1] is a paper model of the basic objects and relationships of a department, section, system, business unit, or company. The model illustrates essential data groups (i.e., objects) and how those objects are related[2] by removing all implementation characteristics or dependencies. An example of "physical" objects that usually don't have a place in an Essential Information Model could be object names such as "Policy History" or "Reconciliation Report." These do have a place in the expanded INFORMATION MODEL, however, since they illustrate implementation.

The Essential Information Model is not at all concerned with the issues outline in AUDIT & SECURITY GUIDELINES. This model, however, must reflect the knowledge and policies of the system at hand.

The only physical objects or nonessential memory that may show up on this model are interfaces with external sources or destinations of data that the DEVELOPMENT TEAM has no control over. Those external sources or destinations may be other systems, departments, companies, or (in certain highly regulated cases) governments.

This should be a simple model when completed.

Who Can Help?

- The DEVELOPMENT TEAM
- CLIENT staff

Who Might Need This?

- The DEVELOPMENT TEAM
- The CLIENT

---

[1] Paul T. Ward discusses the attributes of the "Essential Model" and the "Implementation Model" and others in *Systems Development Without Pain* (New York: Yourdon Press, 1984).

[2] Matt Flavin's *Fundamental Concepts of Information Modeling* (New York: Yourdon Press, 1981) describes basic data modeling as an analytical tool.

What's Inside?

- entity-relationship diagrams (ERDs)
- textual specification for each object
- attributes or data elements associated with each object
- textual specification for each *relationship*

ESSENTIAL PROCESS MODEL

What Is It?

The essence of a system consists of all aspects of the system that *must exist* regardless of the level of technology used to implement the system. The essence of a system consists of:

- its essential *activities* and
- its essential *memory*.

The Essential Process Model[1,2] is a paper model of the basic functions of a new business system, department, section, or business unit. This model illustrates essential process and memory, and will exclude most implementation-*dependent* physical constraints such as edits, audits, technology, time, and transportation. It will not include the handling of any of the issues outlined in AUDIT & SECURITY GUIDELINES.
The only physical constraints, technology, transportation, or predesigned memory that will show up on this model are interfaces with external sources or destinations of data over which the DEVELOPMENT TEAM has no control. Those external sources or destinations may be other systems, departments, or companies.
To accelerate the process of creating this model we can use "event analysis" techniques. This is a fast way of determining the "events" (external and internal) the system must respond to and clearly defining the associated stimuli provided to the system and the planned responses.
Normally, an Essential Process Model is greatly influenced by a high-level ESSENTIAL INFORMATION MODEL. If such an Infor-

---

[1] J. Palmer and S. McMenamin's *Essential Systems Analysis* (New York: Yourdon Press, 1984) differentiates between the "model of essence," which is free of implementation methods, and the "implementation model," which contains all edits, audits, controls, transportation, time, and technology.

[2] Paul T. Ward discusses the attributes of the "Essential Model" and the "Implementation Model" and others in *Systems Development Without Pain* (New York: Yourdon Press, 1984).

mation Model is not produced, it will take longer to produce a good Essential Process Model.

This should be a simple model when completed.

Who Can Help?

- The DEVELOPMENT TEAM
- The CLIENT

Who Might Need This?

- The DEVELOPMENT TEAM
- The CLIENT

What's Inside?

- list of system "events" (external and internal)
- data flow diagrams
- a high-level narrative for each process primitive
- a DATA DICTIONARY

FUNCTIONAL SYSTEM MODEL

What Is It?

The Functional System Model is simply the "packaging" of the INFORMATION MODEL and the PROCESS MODEL. The data dictionaries for each of these models are combined into one DATA DICTIONARY.

Who Can Help?

- The DEVELOPMENT TEAM
- The CLIENT's staff
- The Data Dictionary administrator
- The department's secretary

Who Might Need This?

- The DEVELOPMENT TEAM
- The QUALITY MANAGEMENT TEAM
- The CLIENT

- Internal Audit
- Systems Security

What's Inside?

- INFORMATION MODEL
- PROCESS MODEL
- DATA DICTIONARY (combined, for both preceding models)

FUNCTION TESTING PACKAGE

What Is It?

This is the software and procedures, and their documentation, for testing each function at the lowest (unit) level in the system.

This testing package must define the testing procedures and expected results for both automated and manual procedures.

The definition of the lowest functional level (i.e., primitive) varies a lot from company to company. However, a functions test may consist of:
  the edits
  the processing of a specific transaction
  one calculation, or closely related calculations
  reconciliation of a specific transaction

This deliverable is unique to each version of a system developed.

This testing package is commonly called the "Unit Testing Package."[1]

Who Can Help?

- The QUALITY MANAGEMENT TEAM
- CLIENT staff
- The TESTING TEAM
- Internal Audit
- Systems Security

Who Might Need This?

- The TESTING TEAM

What's Inside?

- test package procedures and software, including
    identification (e.g., a policy or security number)

---

[1] The traditional approach to unit testing is virtually impossible with online systems, and should not be attempted. Instead, dialogues, dialogue segments, and dialogue suites should be tested as "units."

why it is included (with specific reference to the ACCEPTANCE CRITERIA)
expected results
things that have to be done (i.e., a tasklist)

HARDWARE REQUIREMENTS

What Is It?

Defines the new hardware required by the target system.
It isn't necessary to define the existing mainframe, if it is to be used, nor other hardware that is currently available to the client.[1] This deliverable defines only new hardware that needs to be purchased or leased. This is usually produced in conjunction with the purchase of a micro- or personal computer. Rarely does a development team get involved in specifying what kind of mainframe is needed.

Who Can Help?

- The DEVELOPMENT TEAM
- CLIENT management
- DP Hardware Planning group
- Office Systems group
- Corporate Systems Planning

Who Might Need This?

- DP Hardware Planning group
- Office Systems group
- Internal Audit
- Systems Security
- The CLIENT[2]

What's Inside?

- the on-site location of the hardware
- type of hardware (micro, plotter, printer, etc.)
- memory needs and media

---

[1] The OPERATIONS group does, however, need advance notice for scheduling, resource use, etc., when new applications are added to existing workload. This, however, is a project management responsibility.

[2] The CLIENT needs the details of new hardware to authorize the capital expenditure.

- teleprocessing compatibility with other hardware
- required security and audit features
- date required
- vendor and manufacturer
- basic price and amortization schedule

INFORMATION MODEL

What Is It?

The Information Model expands its predecessor ESSENTIAL INFORMATION MODEL into data groups (i.e., objects) and how they relate to each other in the real, imperfect world.

The Information Model[1] is a paper model of all the objects and relationships needed to support the target system. Policies and conventions such as edits, audits, approvals, and controls are a large part of this model of the client's system.[2]

The Information Model, which contains the knowledge to be applied in the system, significantly influences the PROCESS MODEL, which specifies what is done with the data.

All data implementation constraints must show up in this model. Cases such as reports required by governments, and relationships with imperfect database design must be illustrated. Imperfect or redundant relationships (which are often caused by interdepartmental boundary disputes) must be clearly shown. Imperfect relationships and access to source data must be shown.

The text of the Information Model specifies all the rules of the game—that is, the rules, policies, laws, and conventions behind the relationships among objects. The Information Model, however, doesn't necessarily specify exactly how these "rules" should be implemented; that is the job of the PROCESS MODEL.

The Information Model, in terms of policy, must be totally *predictive* when it is complete; that is, it must be able to answer any question concerning what the system must *know* to be able to function.

It is not usually a simple model when completed.

---

[1] Paul T. Ward discusses the attributes of the various system models in *Systems Development Without Pain* (New York: Yourdon Press, 1984).

[2] Matt Flavin's *Fundamental Concepts of Information Modeling* (New York: Yourdon Press, 1981) describes basic data modeling as an analytical tool.

Who Can Help?

- The DEVELOPMENT TEAM
- QUALITY MANAGEMENT TEAM members
- The CLIENT
- External interface contacts
- Internal Audit
- Systems Security

Who Might Need This?

- The DEVELOPMENT TEAM
- The QUALITY MANAGEMENT TEAM
- The CLIENT
- Internal Audit
- Systems Security

What's Inside?

- entity-relationship diagrams (ERDs)
- textual specifications for each *object*
- attributes or data elements associated with each object
- textual specification for each *relationship*

INTEGRATION TESTING PACKAGE

What Is It?

This is the software and procedures, and their documentation, for testing the integration of functions in the system.

This testing package must define the testing procedures and expected results for both automated and manual procedures.

The definition of an integration test varies a lot from company to company. However, the integration test may consist of testing:
    combined automated processes
    combined manual processes
    combined manual processes with automated processes

This testing package is incremental as new versions of the system are developed.

Who Can Help?

- The QUALITY MANAGEMENT TEAM
- CLIENT staff
- The TESTING TEAM
- Internal Audit
- Systems Security

Who Might Need This?

- The TESTING TEAM

What's Inside?

- test package procedures and software, including
    identification (e.g., a policy or security number)
    why it is included (with specific reference to the ACCEPTANCE CRITERIA)
    expected results
    things that have to be done (i.e., a tasklist)

INTERSYSTEM TESTING PACKAGE

What Is It?

This is the software and procedures, and their documentation, for testing the interfaces between the target system and other systems.

This testing package must define the testing procedures and expected results for both automated and manual procedures.

The definition of an intersystem test varies a lot from company to company. It is often called a "Release Test." However, the intersystem test may consist of testing:
    accepting data from other systems
    providing data to other systems
In both cases, the edits, audits, and controls must be thoroughly tested as per the ACCEPTANCE CRITERIA.

This test package is incremental as future versions of the system are developed and released.

Who Can Help?

- The QUALITY MANAGEMENT TEAM
- CLIENT staff
- The TESTING TEAM

- Internal Audit
- Systems Security

Who Might Need This?

- The TESTING TEAM
- Members of other development and support teams

What's Inside?

- test package procedures and software, including
    identification (e.g., a policy or security number)
    why it is included (with specific reference to the ACCEPTANCE CRITERIA)
    expected results
    things that have to be done (i.e., a tasklist)

MODULE SPECS

What Is It?

A specification of what a program module must do.
". . . a good specification ought to be clear, unambiguous, concise (to avoid excessive volume), and complete (so that no essential element is left unspecified)."[1]
Also, "Each mini-spec must describe rules governing transformation of data flows arriving at the associated primitive into data flows leaving it."[2]

Who Can Help?

- QUALITY MANAGEMENT TEAM staff

WHO MIGHT NEED THIS?

- The QUALITY MANAGEMENT TEAM

What's Inside?

- a process specification that is a subset of everyday language, with limited syntax, limited vocabulary, and an indention convention that calls attention to logical blocking.

[1] From Tom DeMarco's *Structured Analysis and System Specification* (New York: Yourdon Press, 1978), p. 169.
[2] Ibid., p. 170.

- The vocabulary of a Module Spec consists only of:
    imperative verbs
    terms defined in the DATA DICTIONARY
    certain words reserved for logic formulation, and agreed to by the DEVELOPMENT TEAM.

OBJECTIVES & CONSTRAINTS

What Is It?

The scope of the target system, as seen by the CLIENT[1] including the priorities, schedules and deadlines, and budget allocations. It should include what will *not* be done by the DEVELOPMENT TEAM.

Who Can Help?

- CLIENT management
- The DEVELOPMENT TEAM

Who Might Need This?

- CLIENT management
- The DEVELOPMENT TEAM
- Internal Audit
- Systems Security

What's Inside?

- a definition of project scope, including assumptions
- objectives and priorities (depends on VERSION PLAN)
- constraints, in terms of
    policy (company and government)
    technical (hardware and software)
    operational (personnel, etc.)
    financial
    scheduling and target dates
    data privacy
- a definition of system criticality, usually resulting from risk analysis provided by CLIENT staff.

[1] Although this document should be prepared from the viewpoint of the client, it is actually written most often by an analyst from the data processing department. Setting the tone, then, is vitally important. Lois Zells has written a practical, useful article entitled A Practical Approach to a Project Expectations Document, *Computerworld*, August 29, 1983. This report can be very useful to experienced or novice project managers when planning the preparation of a project's Objectives & Constraints.

OFFICE EQUIPMENT REQUIREMENTS

What Is It?

A definition of all the office equipment and furniture needed to support the new system and its people, once implemented, which is not currently in place.

Who Can Help?

- The DEVELOPMENT TEAM
- CLIENT staff
- Office Systems Group
- Office Equipment & Facilities Planning group

Who Might Need This?

- The CLIENT[1]

What's Inside?

- a complete description of the equipment
- what the equipment will be used for
- where the equipment will be located
- when it will be needed by
- vendors/manufacturers
- basic price, possibly with amortization schedules

OPERATIONS

What Is It?

That part of the company or organization responsible for running the system once it is "in production." Specific responsibilities of Operations can include:

- the scheduling of production jobs
- the monitoring of production jobs while they run
- the distribution of output

Operations is also usually charged with the maintenance of the hardware environment. As such, they may control:

---

[1] The client might need this information in order to authorize expenditures for office equipment.

- access to the machine, data storage devices, and so on
- operating and systems software
- teleprocessing functions
- system availability and the like

The smaller the company, the more integrated Operations may be with the CLIENT's area. In the extreme case of a small application running on a microcomputer, there may be no Operations group at all, but there will still be people bearing the Operations responsibility.

OPERATIONS FLOWCHART

What Is It?

This is a conventional diagram of the automated processes of the target system showing its input, output, jobs, job steps, intermediate files, physical devices (processors), subsystem groupings, programs, and control requirements. It's sometimes called a "systems flowchart." It is *not* a logic flowchart.

Who Can Help?

- QUALITY MANAGEMENT TEAM staff

Who Might Need This?

- OPERATIONS group
- QUALITY MANAGEMENT TEAM staff

What's Inside?

a graphic representation (i.e., drawing) of the system, including the following:

- programs
- transactions
- data stores
- system interfaces
- system inputs
- system outputs
- scheduling information

OPERATIONS SUPPORT PROCEDURES

What Is It?

A document defining all the things that need defining to operate the system, respond to its quirks, how to recover in the event of disaster, when it should run, and who to call when there's a problem. It's pretty well standard stuff.

These procedures are incremental as future versions of the system are created and delivered.

Who Can Help?

- The QUALITY MANAGEMENT TEAM
- OPERATIONS group
- Internal Audit
- Systems Security

Who Might Need This?

- OPERATIONS group

What's Inside?

- OPERATIONS FLOWCHART
- DISASTER RECOVERY PLAN
- restart/recovery procedures[1]
- run instructions
- security notice and procedures
- scheduling requirements
- problem identification and recovery procedures
- service level reporting procedures[2]
- response monitoring procedures
- action/reaction procedures
- escalation procedures
- control checking procedures
- distribution procedures
- data entry procedures

---

[1] Louie A. Molnar has written an exceptionally good article, entitled *Recovery Control Standards* (*EDP Auditor*, Fall 1983, pp. 11–18), which can be used as a basis for developing backup, restart, recovery, and control/security standards in just about any environment.

[2] Service level requirements for a system are defined in ENVIRONMENTAL ACCEPTANCE CRITERIA. It is a definition of how the system will behave in a specific environment.

- input files specifications
- output files specifications
- file retention rules
- hardcopy instructions
- microfiche instructions
- nonstandard billing procedures
- call list
- other special procedures

OPERATIONS TRAINING PACKAGE

What Is It?

A "package" used to train OPERATIONS staff in the support and running of the new system. This kind of training program is usually only produced with the more complex and sophisticated systems. Most training offered to OPERATIONS happens rather informally during the testing process, since most conventional systems have the same basic rules for operation. In any event, operators have been known to perform miracles when there have been no run instructions whatsoever. They also rely on the JCL a lot.

This training package is incremental as new versions of the system are developed.

Who Can Help?

- The QUALITY MANAGEMENT TEAM

Who Might Need This?

- OPERATIONS group

What's Inside?

- how to use the OPERATIONS SUPPORT PROCEDURES
- an overview of the system, including operational interfaces with other systems, via the OPERATIONS FLOWCHART
- detailed study of the system's JCL
- schedules and priorities of the system
- operational audit methods
- security procedures of the system
- data handling procedures

- data entry procedures
- disposition of data stores and reports
- backup and recovery procedures
- detailed study of the DISASTER RECOVERY procedures

PERSONNEL REQUIREMENTS

What Is It?

Defines the staffing needs of the system under development, after it is operational. Personnel requirements include any staff needed by the CLIENT and any departments affected by the change in the CLIENT's operation.

In most cases the detailed work of deciding how many people should be hired (or outplaced), who they will report to, how much they will make, who has to be retrained—and how this affects the COST-BENEFIT STATEMENT—will be done by CLIENT management. This should never be done by a DEVELOPMENT TEAM, although contemporary literature would suggest that the DEVELOPMENT TEAM does just about everything!

Who Can Help?

- The DEVELOPMENT TEAM
- CLIENT management
- human resources department

Who Might Need This?

- CLIENT management

What's Inside?

- the new functions needing people
- skills and attributes of those people

PRELIMINARY AGREEMENT

What Is It?

The Preliminary Agreement is simply the packaging, sometimes spiffy, sometimes not, of what has been agreed by the DEVELOPMENT

TEAM and CLIENT management. This agreement relates only to the fact that an ESSENTIAL BUSINESS MODEL is to be created or refined and recommendations made to the CLIENT concerning the possible solution to the problem identified in the first place.

Who Can Help?

- You don't need a lot of help for this one!

Who Might Need This?

- CLIENT management
- The DEVELOPMENT TEAM
- Internal Audit
- Systems Security

What's Inside?

- OBJECTIVES & CONSTRAINTS
- APPROVAL GROUP LIST
- CLIENT LIST

PROCESS MODEL

What Is It?

The Process Model expands the ESSENTIAL PROCESS MODEL into a real-world specification statement. It does not distinguish between automated and manual processes, nor between the various automated processors (online, batch, etc.) that are possible for the system.

The Process Model[1,2] is a paper model of all the functions identified to support internal and external systems events. Physical processes such as edits, audits, controls, technology, time, and transportation between processes *are a major part* of this model of the system.

The Process Model (*what is done with the data*) is usually significantly

---

[1] The data flow diagram as described by Tom DeMarco in *Structured Analysis and System Specification* (New York: Yourdon Press, 1978) is the basis for the Process Model.

[2] Paul T. Ward discusses the attributes of the various system models in *Systems Development Without Pain* (New York: Yourdon Press, 1984).

influenced by the INFORMATION MODEL (*which contains the knowledge to be applied in the system*).[3]

All specific physical constraints must show up in this model. These include: (1) technological decisions (e.g., the decision to purchase the famous DataSchwartz 2938 computer, which only handles punched cards); (2) imperfect or redundant methods of transportation (which can be caused by interdepartmental boundary disputes); (3) external constraints, such as a specific method of collecting data (an example is when the system is obliged to get data from the General Accounting system when the ESSENTIAL PROCESS MODEL shows it should clearly come from the Inventory Control system); (4) any additional external or policy constraints, such as the antiquated technology imposed on your system by the bank you deal with.

The Process Model illustrates and specifies all system planned responses, with the stimuli and data essential to the processes that deliver the planned responses. It includes the edits, audits, controls, time, and transportation processes the system needs, prior to the allocation of processor boundaries. AUDIT & SECURITY GUIDELINES mention things to think about.

The Process Model must be totally *predictive* when it is complete; that is, it must be able to answer any question concerning its behavior in a specific environment. It is *not* usually a simple model when completed.

Who Can Help?

- The DEVELOPMENT TEAM
- QUALITY MANAGEMENT TEAM members
- The CLIENT
- External interface contacts
- Internal Audit
- Systems Security

Who Might Need This?

- The DEVELOPMENT TEAM
- The CLIENT
- Internal Audit
- Systems Security

[3] "It's easy to see the practical usefulness of a process schema, because it serves as a blueprint for setting up a system and shows the pathways that must be established for transporting data so that processing can occur. In the same way, a data schema serves as a blueprint for cross-referencing stored data so that it can be retrieved to answer questions." Ibid., p. 91.

What's Inside?

- data flow diagrams
- a narrative for each process primitive
- a DATA DICTIONARY

PROCESSOR CONFIGURATION MODEL

What Is It?

The Processor Configuration Model is a copy of the PROCESS MODEL, with the automation boundary and all interprocessor boundaries marked on it. The interprocessor boundaries are drawn such that for each subsystem identified, all functions will be:

- implemented on the same processor
- executed in the same time frame (i.e., online, daily, weekly)

After such partitioning, the design and building of each subsystem can be considered separate tasks and allocated to subsets of the DEVELOPMENT TEAM, if resources permit.

The Processor Configuration Model is a working document, and the actual implementation may change slightly as design and construction continue. It is retained, however, so hardware and temporal architecture can be identified easily on one data flow diagram.

Who Can Help?

- CLIENT management
- QUALITY MANAGEMENT TEAM staff
- OPERATIONS group

Who Might Need This?

- CLIENT management
- The QUALITY MANAGEMENT TEAM
- The TESTING TEAM
- The DEVELOPMENT TEAM
- Corporate Systems Planning

What's Inside?

- carved up data flow diagrams

QUALITY MANAGEMENT TEAM (THE)

What Is It?

The team of people associated with the implementation and quality management of the client's system. This team also feeds and cares for the system after implementation. They have previously been known as "maintenance programmers," but this is far too limiting a role.

In some companies it is possible that the DEVELOPMENT TEAM and the Quality Management Team consist of the same people, or at least have some crossover.

The roles of the Quality Management Team could be to:

- prevent degradation of a system's "goodness"
- maintain and give a conscience to Quality Standards
- measure quality
- participate in postproject reviews of results and methods used to produce the results
- evaluate the techniques in use
- improve quality accounting methods
- establish and publicize Quality Standards (as guidelines)
- specialize in current technology and how it can be used to solve business problems
- develop and maintain expertise in
    documentation
    database
    telecommunications
    testing
    programming
- be responsible for implementation of systems
- control the SYSTEM IN PRODUCTION
- be the CLIENT liaison re: production
- be advisers to DEVELOPMENT TEAMs on technical matters
- have an active role in Quality Assurance on development projects

The team should never dictate methods, measure people, or standardize any untried techniques.

Attributes:

Although these people are much more technical and oriented toward use of the left hemisphere of the brain, they have basically the same attributes as DEVELOPMENT TEAM people.

REQUEST FOR SERVICE

What Is It?

Any identified need for new functions for a proposed system, or a change to an existing system. This could simply be the client stating that a problem exists and that possible solutions should be evaluated. It could also be the result of monitoring a system during the shakedown period after it is moved to production.

Who Can Help?

- CLIENT management
- The QUALITY MANAGEMENT TEAM

Who Might Need This?

- The DEVELOPMENT TEAM
- The QUALITY MANAGEMENT TEAM

What's Inside?

Whenever possible:

- a definition of the existing or potential problem
- a definition of the client's perception of the response to the problem
- special circumstances, such as required interfaces with other departments, systems, or companies
- budget and scheduling constraints
- supporting documentation from DOCUMENTATION HEAVEN

SOFTWARE SUPPORT REQUIREMENTS

What Is It?

Specifies the *systems/support* software and *application* software that can be bought to support the target system in its operating environment.

Who Can Help?

- The QUALITY MANAGEMENT TEAM
- DP Software Planning group
- Office Systems group

Chap. 9   State Transition Diagrams   251

Who Might Need This?

- DP Software Planning group
- Office Systems group
- The CLIENT[1]
- Internal Audit
- Systems Security

What's Inside?

- name of software and its generic type
- what it will do in the target system
- access methods used and how this conforms (or not) with the system
- compatibility with the general environment
- security and audit features of the software
- enhancements needed to the software to meet the requirements of the FUNCTIONAL SYSTEM MODEL
- date needed
- vendor and manufacturer
- basic price

STATE TRANSITION DIAGRAMS

What Is It?

As we use them, State Transition Diagrams (STDs) are Mealy model finite automata without output. (Aren't you glad you looked this up?) Each state represents a certain type of externally observable behavior (a status); the state is passive, and is named after the behavior of the system. Transitions occur at points in time, and cause the system to change state. Associated with each transition is a condition which describes under what circumstances the change of state will occur, and a set of actions (possibly null) which will occur at the time of change of state.

Who Can Help?

- Ward and Mellor[2]

---

[1] The client may need the details to authorize the funds for the software investment.

[2] Paul T. Ward and Stephen J. Mellor provide several good examples of State Transition Diagrams in *Structured Development for Real-Time Systems* (New York: Yourdon Press, 1985). It should be noted, however, that the authors have found State Transition Diagrams to

Who Might Need This?

- Designers of real-time or process control systems.
- STDs might occasionally be needed by online dialogue designers to model complex, non-hierarchical dialogues.

What's Inside?

- The diagram itself with all states, transition conditions and transition actions labeled. There is no further breakdown—the STD itself can serve as the specification for a process control bubble in a DFD.

STRUCTURE CHARTS

What Is It?

A hierarchical model of the processes of a subsystem showing all the functions (modules) and the data passed to and from each module.

Who Can Help?

- QUALITY MANAGEMENT TEAM staff
- Technical design experts
- Meilir Page-Jones[1]

Who Might Need This?

- The QUALITY MANAGEMENT TEAM
- The TESTING TEAM
- The DEVELOPMENT TEAM

What's Inside?

- diagrams and narrative, associated with a DATA DICTIONARY

STUFF KEPT AND MAINTAINED

What Is It?

This is all the "documentation" produced by the DEVELOPMENT and QUALITY MANAGEMENT TEAMs during the design and

---

be far too powerful a tool for commercial online and interactive systems, and more useful in a process control environment.

[1] Everything you always wanted to know (just about) is contained in Meilir Page-Jones's *The Practical Guide to Structured Systems Design* (New York: Yourdon Press, 1980).

implementation of the target system. All of it is kept in DOCUMENTATION HEAVEN.

Coping with "Versionity"

Since most systems are developed in versions or "releases" (i.e., several projects to be integrated), documentation will be in a state of flux. To avoid confusion, it is essential that the documents relating to different system releases be kept separate.

This is analogous to separating production and test code. As soon as the first version of a system is implemented, it becomes the production version, and the models for it become the production documentation. They must reflect the system *as it is*, not as it was when we were developing it or as it will be in the future.

So production models are changed—and must be changed—only to reflect modifications to the current version of the system. They are not changed to reflect the next version of the system *until* the next version is implemented.

Separate the Versions

Therefore, put away all production models (*Essential Business Model, Functional System Model, Design Blueprint*). File diagrams and text in suitable folders in your sublibrary. Move to Production Source all the TSO datasets containing text for narratives, mini-specs, data dictionaries, and so on.

You can make copies (electronic or paper, whichever is appropriate) of these models to use as a starting point for working on the next version of the system. The new documents developed during this process will eventually replace corresponding parts of the production documentation.

Possibility of Change

Since the *Essential Business Model* is unlikely to change much, you probably won't need to supersede a "previous version" when the implementation of the next version takes place.

The *Functional System Model* (FSM) will change, mainly as we integrate new versions into a larger cohesive system. It is possible to have three versions of an FSM: one for the current system in production, one for the next version of the system to be implemented, and a *mañana* one for long-term goals. If this is the case, keep the *mañana* FSM in the project leader's desk drawer—don't let it be mistaken for either the current version of the system or the next version.

The impact of change will, of course, be greatest on the *Design Blueprint*. Make sure that the revised model reflects the changes you had to make to the model of version n. One way to do this is to record the changes on the copies of the *Design Blueprint* documents that you are using to work on new versions.

Aliases?

- "THE STUFF"
- "STUFF"
- Rarely referred to as "the right stuff"

Who Might Need This?

- The QUALITY MANAGEMENT TEAM[1]
- A future DEVELOPMENT TEAM[2]
- The CLIENT's staff[3]
- The Keeper of Good Stuff[4]

What's Inside?

• ESSENTIAL PROCESS MODEL • ESSENTIAL INFORMATION MODEL	ESSENTIAL BUSINESS MODEL
• PROCESS MODEL • INFORMATION MODEL	FUNCTIONAL SYSTEM MODEL
• STATE TRANSITION DIAGRAMS • STRUCTURE CHARTS • MODULE SPECS • DATA DICTIONARY	DESIGN BLUEPRINT

- PRELIMINARY AGREEMENT[5]
- COST-BENEFIT STATEMENT[5]
- VERSION PLAN[5]
- TEST PLAN[6]
- ACCEPTANCE CRITERIA[5]
- PROCESSOR CONFIGURATION MODEL

[1] They are going to maintain the system, at least the part inside the automation boundary!

[2] If, during maintenance, the ESSENTIAL BUSINESS MODEL or FUNCTIONAL SYSTEM MODEL needs to be modified, a DEVELOPMENT TEAM will again be involved.

[3] The client's staff have equal access to DOCUMENTATION HEAVEN since: (1) it's their system; (2) they should continually be upgrading the INFORMATION MODEL; (3) they should maintain the part of the PROCESS MODEL that's outside the automation boundary; (4) many of them were part of the DEVELOPMENT TEAM and will need to refer to it (because of footnote 2).

[4] Who else would surround it with Goodness and Light?

[5] Strictly for historical purposes. This isn't maintained.

[6] Kept and maintained specifically for future releases of the system so this work will not have to be repeated.

- DATABASE DESIGN
- OPERATIONS SUPPORT PROCEDURES
- TEST PACKAGES
- CODED PROGRAMS
- System History, such as a "Change Log"

SUPERSET SYSTEM DEVELOPMENT MODEL

What Is It?

- A model of all processes potentially involved in the development of a business system, whether manual, automated, or a combination thereof, from which a subset can be drawn to suitably describe any specific system's development process.
- A generic systems life cycle model showing the functions that will occur and the deliverables that must be produced in order to create the universal system.
- A strategy, or *game plan for systems development* that suggests that each system's development process will be unique, and that a unique model must be created for each system to be developed prior to its development.
- A book by Frantzen and McEvoy.

Who Might Need This?

- Users
- Project managers
- Data processing professionals
- Business systems analysts
- Those with questions to be answered
- Those with answers to be questioned

What's Inside?

- data flow diagrams
- process descriptions
- a dictionary defining potential deliverables
- years of experience
- wry humor
- sage advice
- heartwarming assurance that it is possible to deliver systems that the user wants, on time, and at the right price

SYSTEM IN PRODUCTION

What Is It?

The SYSTEM RELEASE once it has been moved to production; the existing version of a system for which changes are being considered.

What's Inside?

- software and hardware and anything else needed for a system in production

SYSTEM RELEASE

What Is It?

The latest version of the system, fully tested and acceptable to the CLIENT (based on ACCEPTANCE CRITERIA), but prior to the move to production.

Each version of a system, after integration with the prior release becomes a new System Release. All independent versions developed must maintain the incremental integrity of previous releases.

Who Can Help?

- The TESTING TEAM
- QUALITY MANAGEMENT TEAM members
- The DEVELOPMENT TEAM
- Internal Audit
- Systems Security

What's Inside?

- the system's software

SYSTEMS SUPPORT REQUIREMENTS

What Is It?

The Systems Support Requirements is simply the "packaging" of the various "physical" things needed to build and operate the new system.

Who Can Help?

- The QUALITY MANAGEMENT TEAM
- OPERATIONS group or Technical Support
- Office Systems group
- Forms Design group

Who Might Need This?
- CLIENT staff members
- OPERATIONS group or Technical Support
- Office Systems group
- Forms Design group
- Internal Audit
- Systems Security

What's Inside?
- HARDWARE REQUIREMENTS
- SOFTWARE SUPPORT REQUIREMENTS
- PERSONNEL REQUIREMENTS
- OFFICE EQUIPMENT REQUIREMENTS
- DOCUMENT REQUIREMENTS

TEST FILES

What Is It?
These are the files containing test data that are needed to test the system. Test Files should be incremental as future versions of the system are to be tested.

Who Can Help?
- The QUALITY MANAGEMENT TEAM
- The TESTING TEAM
- CLIENT staff

Who Might Need This?
- The TESTING TEAM

What's Inside?
- Test data
- DATABASE DESIGN for TEST FILES

TEST PACKAGES

What Is It?
These are the various testing packages needed to ensure the system meets the ACCEPTANCE CRITERIA. The packages are based on the TEST PLAN for each system. All Test Packages should be incremental

to handle regression testing when releasing future versions of the system.

Who Can Help?

- The QUALITY MANAGEMENT TEAM
- OPERATIONS group
- CLIENT staff
- The TESTING TEAM
- Internal Audit
- Systems Security

Who Might Need This?

- The TESTING TEAM

What's Inside?

- FUNCTION TESTING PACKAGE
- ENVIRONMENT TESTING PACKAGE
- INTEGRATION TESTING PACKAGE
- INTERSYSTEM TESTING PACKAGE

TEST PLAN

What Is It?

A document, specifying testing objectives and how to attain them, for both automated and manual system functions. It is substantially based on the ACCEPTANCE CRITERIA.

Although it is common practice for data processing staff to develop the TEST PLAN, it's our belief that the people primarily responsible for its development should be the CLIENT's staff (who are already on the DEVELOPMENT TEAM).

The Test Plan is incremental. As future versions of the system are developed, the Test Plan should be expanded to include regression testing of previous system releases.

Who Can Help?

- The QUALITY MANAGEMENT TEAM
- OPERATIONS group
- CLIENT staff

Chap. 9 Tested Conversion System

- The TESTING TEAM[1]
- Internal Audit
- Systems Security

Who Might Need This?

- The TESTING TEAM
- Internal Audit
- Systems Security

What's Inside?

- testing objectives (based on and with specific reference to ACCEPTANCE CRITERIA
- a definition of the approval process, the schedule, and who is involved in it
- a definition of the testing stages needed, so the TEST PACKAGES can be built
- task and responsibility lists
- resources needed (people and machines) from the different parties and departments involved in testing
- an estimate of how long it will take and how much it will cost

TESTED CONVERSION SYSTEM

What Is It?

The conversion system's software, fully tested and acceptable to the CLIENT (based on [Conversion] ACCEPTANCE CRITERIA), but prior to the actual database conversion.

Who Can Help?

- The TESTING TEAM
- QUALITY MANAGEMENT TEAM members
- The DEVELOPMENT TEAM
- Internal Audit
- Systems Security

---

[1] Strictly as internal consultants. The Testing Team should *never* directly create a Test Plan, since they will be executing it.

Who Might Need This?

- The DEVELOPMENT TEAM

What's Inside?

- The conversion system's software

TESTING TEAM (THE)

What Is It?

The team of people associated with the Dark Side of the Force, who do the testing and therefore the quality assurance of a client's system. Ideally, these people should be kept away from all development or maintenance to avoid any conflict of interest when they engage in testing. Also, their role should be one of *destructive testing*, so that the idea of destroying bugs (rather than just discovering them or putting them elsewhere) is continually fostered.

In some companies, however, it is possible that the DEVELOPMENT TEAM and/or the QUALITY MANAGEMENT TEAM and/or the Testing Team consist of the same people, or at least have some crossover.

The (Destructive) Testing Team should be issued black jeans and shirts (with "DTT" on the front and back) and a black cape to signify their special status. They should be fed steak tartare as often as possible. Their sole mission in life should be to subvert the system. Any system. It is not necessary to pay them well; destructive testers do this kind of thing and enjoy it at the same time. Any pay received is a bonus.

The roles of the Testing Team could be to:

- prevent degradation of a system's "goodness" by removing all the defects they can find[1]
- measure defect quantity, location, and so on
- participate in postproject reviews of results and methods used to produce results, specifically reviews of testing techniques used
- evaluate the testing techniques in use and available through vendors and other industry sources
- improve defect accounting methods
- be advisers to DEVELOPMENT TEAMS in development of TEST PLANs

This team should never measure people.

[1] Tom DeMarco, in *Controlling Software Projects: Management, Measurement & Estimation* (New York: Yourdon Press, 1984), discusses software quality and associated metrics, including defect removal.

Attributes:

They should be very suspicious by nature, believing nothing told them by the DEVELOPMENT TEAM or the QUALITY MANAGEMENT TEAM (if they are different people); have a love for asking questions and creating "what-if" scenarios. Their heart's desire should be to find a new way of subverting the system. Their thinking should be extremely analytical (left hemisphere of the brain), but with a lot of creativity thrown in (right hemisphere of the brain). They should be the kind of people who laugh when they discover a defect. Being ambicephalic, they enjoy double headers.

USER DOCUMENTATION

What Is It?

A document that describes all user activities that are part of the system. It is based on the PROCESS MODEL and the DIALOGUE DESIGN. It should be succinct and not encumbered with irrelevant details of the *internals* of the system.[1] Responsibilities are also defined. User Documentation is incremental as future versions of the system are developed and released. Ultimately, User Documentation for the online component of a system should take the form of *help* screens.

Who Can Help?

- QUALITY MANAGEMENT TEAM staff
- CLIENT staff

Who Might Need This?

- CLIENT user staff

What's Inside?

- defines the various activities carried out and their purpose
- identifies input data and the method of obtaining the data
- defines how the data should be organized and how they should be checked for correctness and completeness

---

[1] The CLIENT's clerical staff should be well trained (see CLIENT TRAINING PACKAGE) and able to understand and do their work. But since most people don't carry a training package with them forever and certainly won't look at the thirty-seven acres of shelving needed for the system's documentation, this document should simply be a reminder of "most common things" or a series of "cheat sheets." Everyone should have a personal copy; things that sit on a shelf or in a library don't get used very often. This deliverable is not intended to be very big or complicated.

- explains any rules, policies, and formulas that are applied in dealing with the data
- identifies the output data and how to obtain them
- explains anything that can go wrong and how to deal with it when it does
- provides "cheat sheets" to explain the interface dialogue with machines
- specifies the schedule of events
- An example of typical contents follows:
- General Information
  - what information/manuals/tables are available where
  - how to use/find things in the reference manual(s)
  - what is the "cheat sheet"
  - how to ask for system changes
  - who to call in case of difficulty
  - how to report a problem

- Terminal Available to User Staff
  - how to logon to MaxiNet
  - what to do and who to call if terminal fails
  - how and when to reorder supplies

- Screen Dialogues
  - how to work with the default screen flow
  - how to get from screen to screen outside the defaults
  - how to get data (inquire)
  - how to manipulate data
  - what security features there are

- Backup and Recovery
  - what happens, when
  - how long it takes to happen
  - how data integrity is maintained

- Functions
  - how to prepare data for the system
  - how to get data into the system
  - what to do with data coming out of the system
  - when data is no longer available
  - what is on microfiche and what to do with it, when
  - what to do when the automated system is unavailable

- Error Handling
  - how to know that there are errors
  - what to do about errors
  - what to do about missing data
  - what to do about just plain wrong data
  - what to do when the system just won't go
  - who to call when nothing else works

VERSION PLAN

What Is It?

A partitioning of the functions to be delivered to the client, such that each segment can be implemented independently, gradually giving the client an increasing level of support until the entire system is installed. It's based on the PROCESS MODEL and the PROCESSOR CONFIGURATION MODEL.

Who Can Help?

- The DEVELOPMENT TEAM
- CLIENT management

Who Might Need This?

- CLIENT management
- The DEVELOPMENT TEAM

What's Inside?

Each segment, or version, will have:

- a description of the functions included
- an itemization of the work to be done (e.g., enhance Functional System Model, expand Design Blueprint)
- a target delivery date, preferably for each major deliverable
- probable resource loading and cost
- and, usually, a name or number (e.g., System Version 2, System Version 2.1)[1]

---

[1] A Version Plan can be used as input to Objectives & Constraints when work on a new version of a system is started.

CHAPTER 10

LIST OF THE DIAGRAMS OF THE SUPERSET MODEL

For the sake of easy reference, a complete list of the diagrams and illustrations appearing in this book follows, with the corresponding page number of where the diagram can be found. Those diagrams comprising the Frantzen/McEvoy SuperSet System Development Model are shown in boldface type.

Fig. No.	Diagram Name	Page No.
1	Context: Essence of Build Client System	18
2	Essence of Build Client System—The Development Life Cycle	19
3	The Client Group, Users, and Development Team	21
3.1	The Perfect Team	22
3.2	A Workable Team	22
3.3	The Impossible Team	22
4	Context of the Process to Prove Fermat's Last Theorem	24
5	Stuff Going on in the Background	25

List of the Diagrams of the Superset Model

Fig. No.	Diagram Name	Page No.
6	Diagram 0 of the SuperSet System Development Model	28
7	**Development Life Cycle Context Diagram**	**31**
8	**Diagram 0 of the SuperSet System Development Model**	**35**
9	Neighborhood Diagram—Set Up Preliminary Agreement	36
10	Neighborhood Diagram—Create Essential Business Model	37
11	Stuff Going on in the Background	40
12	Neighborhood Diagram—Model System Functions	44
13	Neighborhood Diagram—Design the System	47
14	Neighborhood Diagram—Build the System	49
15	Neighborhood Diagram—Convert Current Data	51
16	Neighborhood Diagram—Develop Training Materials	52
17	Neighborhood Diagram—Implement System	54
18	Neighborhood Diagram—Maintain the System	55
19	**Set up Preliminary Agreement (1)**	**61**
20	Create Models of Essence (2.1)	64
21	**Create Essential Business Model (2)**	**67**
22	**Create Models of Essence (2.1)**	**70**
23	Record Hire	72
24	MaxiChange Currency Exchange System—Entity-Relationship Diagram, Part 1	75
25	MaxiChange Currency Exchange System—Entity-Relationship Diagram, Part 2	76
26	Customer Wishes to Exchange Currency—Event Model	79
27	Customer Wishes to Exchange Currency, Revised—Event Model	80
28	Time to Replenish Currency—Event Model	81
29	Time to Deposit Excess Currency—Event Model	82
30	Time to Receive Interest—Event Model	82
31	Time to Pay Vendor Bank Charges—Event Model	83
32	Time to Pay Rent—Event Model	84
33	Time to Pay Employees—Event Model	84
34	Time to Pay Other Expenses—Event Model	85
35	Market Exchange Rates Change—Event Model	86

List of the Diagrams of the Superset Model Chap. 10

Fig. No.	Diagram Name	Page No.
36	Time to Recognize New Currency—Event Model	87
37	Time to De-recognize Currency—Event Model	88
38	Time to De-recognize Currency—Event Model A Second View	89
39	MaxiChange Currency Exchange System—Entity-Relationship Diagram, Part 1	90
40	Partially Amended Entity-Relationship Diagram—MaxiChange Currency Exchange System	91
41	Create Models of Essence (2.1)—Partial	92
42	Government Wants Employment Reports—Event Model	93
43	Government Wants Payroll Remittances—Event Model	94
44	Time to Report Profit-Loss to Head Office—Event Model	95
45	Context Diagram—MaxiChange Financial System	96
46	Diagram 0—MaxiChange Financial System	97
47	Exchange Currency (1)	98
48	Maintain Currency Inventory (2)	99
49	Pay Expenses (3)	100
50	Issue Payroll (4)	101
51	Issue Head Office Information (5)	102
52	Refine Objectives & Constraints (2.2)	103
53	Prepare Cost-Benefit (2.4)—Level 1	104
54	**Prepare Cost-Benefit (2.4)—Level 2**	**105**
55	Consolidate Essential Business Model (2.3)	110
56	EPM for an Inventory Management System	111
57	An EPM SuperModel of Three Departments	112
58	**Model System Functions (3)**	**113**
59	Entity-Relationship Diagram of a Human Resources System	115
60	Data Dictionary Entry for Relationship "Records"	116
61	Data Dictionary Entry for Object "Employee"	117
62	Expanding the Essential Models with Physical Control Functions	120
63	Sample Data Flow Diagram from a Human Resources System	124
64	Data Dictionary Entry—Insufficient Net Pay	125

Sec. List of the Diagrams of the Superset Model 267

Fig. No.	Diagram Name	Page No.
65	Sample Plain-Language Process Specification	125
66	**Revise Cost-Benefit (3.4)**	**126**
67	Consolidate the Functional System Model (3.3)	130
68	Record Functional Acceptance Criteria (3.5)	130
69	Human Resources System Acceptance Criteria List	131
70	Cost-Benefit Statement Overview	132
71	Summary: IOIs	134
72	Summary: Custom-Build System	135
73	Summary: Buy Software Package	136
74	**Design the System (4)**	**138**
75	Allocate Functions to Processors (4.1)	139
76	Allocate Activities to Processors	141
77	Design Subset Architecture—Level 1 (4.3)	142
78	**Design Subset Architecture—Level 2 (4.3)**	**143**
79	**Build the System (5)**	**148**
80	Testing the System	149
81	Create User Documentation (5.7)	151
82	**Convert Current Data (6)**	**153**
83	**Develop Training Materials (7)**	**154**
84	**Implement the System (8)**	**156**
85	Evaluate Project Results	158
86	The Typical Systems Life Cycle	159
87	The Complete Systems Life Cycle	160
88	Traditional View of Systems Maintenance	160
89	**Maintain the System—Version 1 (9)**	**162**
90	Maintain the System—Version 2	163
91	Diagram 0—SDM for Maintenance	165
92	Diagram 1: Evaluate Request	166
93	Diagram 2: Model Functional Changes	168
94	Diagram 3: Design Program Changes	169
95	Diagram 4: Code and Test	170
96	Diagram 5: Implement Changes	171
97	Core Formal Education	190
98	The Lump Law	208

BIBLIOGRAPHY

ALVAREZ, JOAN, "A Business System Plan," *Journal of Information Management,* IV, no. 2 (1983).

BARNES, DJUNA, *Nightwood* (New York: New Directions Books, 1961).

BATT, ROBERT, "White Collar Crime," *Computerworld,* December 26, 1983, 49–56.

BEELER, JEFFRY, "Timing Said to Be Critical to Training," *Computing Canada,* February 12, 1984.

BLAKENEY, SUSAN, "Computer Crime: A Worldwide Concern," *Computerworld,* December 26, 1983, 57–60.

BLOCK, ROBERT, *The Politics of Projects* (New York: Yourdon Press, 1983).

BOAR, BERNARD H, *Application Prototyping* (New York: John Wiley & Sons, 1984).

BOEHM, BARRY, *Software Engineering Economics* (Englewood Cliffs, N.J.: Prentice-Hall, 1981).

BRILL, ALLAN, *Building Controls into Structured Systems* (New York: Yourdon Press, 1983).

BROOKS, FREDERICK P., Jr., *The Mythical Man-Month,* rev. (Reading, Mass.: Addison-Wesley, 1982).

BUSS, MARTIN D. J., and LYNN M. SALERNO, "Common Sense and Computer Security," *Harvard Business Review,* March–April 1984, 112–121.

CAMPBELL, DIANE E., "The Prototyping Project at Connecticut General," *Journal of Information Management,* IV, no. 1 (1982).

CHRISTENSEN, K., et al., "A Perspective on Software Science," *IBM Systems Journal,* 20, no. 4 (1981).

COUGER, DANIEL J., "In Search of DP Excellence," *Computerworld,* January 16, 1984.

CROSBY, PHILIP B., *Quality Is Free* (New York: New American Library, 1979).

DEMARCO, TOM, *Structured Analysis and System Specifications* (Englewood Cliffs, N.J.: Prentice-Hall, 1979).

———, "Specification Modeling," presented to GUIDE 50, May 5, 1980.

———, *Controlling Software Projects: Management, Measurement & Estimation* (New York: Yourdon Press, 1982).

———, "Future Trends in Software Development," three-day seminar, 1982.

———, "The Issue of Control," *The Yourdon Report,* January–March 1983.

DICKINSON, BRIAN, *Developing Structured Systems: A Methodology Using Structured Techniques* (New York: Yourdon Press, 1981).

DURELL, WILLIAM, "Disorder to Discipline via the Data Dictionary," *Journal of Systems Management,* May 1983, 21–23.

ECKOLS, STEVE, *How to Design and Develop Business Systems* (Fresno, Calif.: Mike Murach & Associates, Inc., 1983).

EDWARDS, BETTY, *Drawings on the Right Side of the Brain* (Los Angeles: J. P. Tarcher, 1979).

FEIGENBAUM, EDWARD A., and PAMELA MCCORDUCK, *The Fifth Generation* (Menlo Park, Calif.: Addison-Wesley, 1983).

FLAVIN, MATT, *Fundamental Concepts of Information Modeling* (New York: Yourdon Press, 1981).

FRANTZEN, TROND, "Productivity 83" (MaxiMoney, Inc., Toronto, March 1983). A paper outlining a productivity improvement program.

———, *Education Program for MaxiMoney Analysts and Designers* (MaxiMoney, Inc., Toronto, October 1983).

———, *The Mentor Program* (MaxiMoney, Inc., Toronto, February 1984).

———, "Better Training Needed for Analysts," *Computing Canada*, September 5, 1985, 17, 23.

———, KEN MCEVOY, and DON DYMENT, *The SuperSet System Development Model* (MaxiMoney, Inc., Toronto, 1983). A 43-minute videotape reviewing the superset concept.

FRANTZEN, TROND, KEN MCEVOY, and DAVID REASON, *The SuperSet System Development Model* (MaxiMoney, Inc., Toronto). Course offered to MaxiMoney employees in 1982, 1983, and 1984.

GANE, CHRIS, and TRISH SARSON, *Structured Systems Analysis: Tools and Techniques* (New York: IST Databooks, 1977).

GILLIN, PAUL, "Outdated Methodology is Seen Crushing DP," *Computing Canada*, April 28, 1983, p. 5.

GLASS, ROBERT L., *Software Reliability Guidebook* (Englewood Cliffs, N.J.: Prentice-Hall, 1979).

GOLEMAN, DANIEL, "The Human-Computer Connection," *Psychology Today*, March 1984, 20–24.

HANSEN, H. DINES, *Up and Running* (New York: Yourdon Press, 1984).

HANSEN, KIRK, *Data Structured Program Design* (Topeka, Kan.: Ken Orr & Associates, Inc., 1984).

HARRIS, LARRY R., "Fifth Generation Foundations," *Datamation*, July 1983, 148–156.

HERTZ, DAVID B., "Artificial Intelligence and the Business Manager," *Computerworld*, October 24, 1983, 19–26.

HILTS, PHILIP J., "The Dean of Artificial Intelligence," *Psychology Today*, January 1983, 28–32.

INMON, BILL, "Rethinking Productivity," *Datamation*, June 15, 1984, 185–188.

JACKSON, MICHAEL A., "Principles of Program Design" (New York: Academic Press, 1975).

———, *Systems Development* (Englewood Cliffs, N.J.: Prentice-Hall, 1983).

JONES, T. CAPERS. *Programming Productivity* (Englewood Cliffs, N.J.: Prentice-Hall, 1986).

KELLY-BOOTLE, STAN, *The Devil's DP Dictionary* (New York: McGraw-Hill, 1981).

KIMMERLY, W. C., "Restricted Vision", *Datamation*, November 1982, 152–160.

KLEIN, MARK M., "The Coming Crisis in MIS: Information System Obsolescence, The Hidden Time Bomb," speech (John Diebold & Associates).

KNUTH, DONALD E., "Structured Programming with Go To Statements" (*ACM Computing Surveys*, December 1974.

LAND, FRANK, "Adapting to Changing User Requirements," *Information & Management*, June 1982, 59–75.

LERNER, ERIC J., "Why Can't a Computer Be More Like a Brain?" High Technology, August 1984, 34–41.

LYBROOK, CHARLES W., "A Positive Approach to Quality Assurance," *Profit Oriented Systems Planning Programs*, March 1982.

MCMENAMIN, STEPHEN M., and JOHN F. PALMER, *The Transition Between Analysis and Design* (New York: Yourdon, Inc., presented at GUIDE 54, Anaheim, Calif., May 14, 1982; rev. June, 1982).

MARTIN, JAMES. *Managing the Data-Base Environment* (Englewood Cliffs, N.J.: Prentice-Hall, 1983).

———, *Essential Systems Analysis* (New York: Yourdon Press, 1984).

MARTIN, TOBY, "The Emperor Has No Clothes," *System Development*, June 1983.

MERRILL, DANIEL, "Productivity Improvement for the 1980's," *Profit Oriented Systems Planning Programs*, December 1981.

MINSKY, MARVIN. *The Society of Mind* (New York: Simon and Schuster, 1986).

MOLNAR, LOUIE A., "Recovery Control Standards," *EDP Auditor*, Fall 1983.

MYERS, GLENFORD J., *The Art of Software Testing* (New York: John Wiley, 1979).

NIES, THOMAS, "How to Significantly Improve Systems Analysis, Design & Maintenance Productivity," *Profit Oriented Systems Planning Programs*, December 1981.

O'CONNOR, JACK, "Made to Last? Never Undervalue the User's Input to System Designs," *Computerworld*, April 25, 1983, 19–28.

ORR, KEN, *Structured Systems Development* (New York: Yourdon Press, 1977).

PAGE-JONES, MEILIR, *The Practical Guide to Structured Systems Design* (New York: Yourdon Press, 1980).

PALMER, JOHN F., "Of Time and Analysis," *Computerworld*, August 13, 1984, 1–12.

PARIKH, GIRISH, and NICHOLAS ZVEGINTZOV, "Managing Stability and Change," *Data Training*, May 1983, 11–12.

PETERS, THOMAS J., and ROBERT H. WATERMAN, JR., *In Search of Excellence* (New York: Harper & Row, 1982).

PROCTOR, LARRY L., "A Cycling Approach to Successful Project Management," *The Yourdon Report*, May–June 1982.

RAMSGARD, WILLIAM C., "Stop Documenting," *Computerworld*, March 1983, 19–23.

RAUDSEPP, EUGENE, "Managing for Superperformance," *Office Administration and Automation*, February 1983.

ROCHE, GERARD R., "Much Ado about Mentors," *Harvard Business Review*, January–February 1979, 14–28.

ROSE, FRANK, *Into the Heart of the Mind* (New York: Harper & Row, 1984).

SCHANK, ROGER C., *The Cognitive Computer* (Reading, Mass.: Addison-Wesley, 1984).

Bibliography

SCHARER, LAURA L., "User Training: Less Is More," *Datamation*, July 1983, 175–182.

SEMPREVIVO, PHILIP C., *Teams in Information Systems Development* (New York: Yourdon Press, 1980).

SILVER, H. FRED, "A Pragmatic Approach to the Management of Software Development Life Cycles," *Proceedings of the 1982 Computer Measurement Group*, International Conference, San Diego, Calif., December 1982, 142–155.

SMITH, WAYNE, "Just What Is Quality Anyway?" *Computerworld*, January 16, 1984, 31–32.

STAFF WRITER, *Quality Assurance Survey Report*, EDPACS, April 1983. This is a summary report of an international survey of the practices and policies of EDP quality assurance.

STAFF WRITER, "Software Engineering and Productivity," *System Development*, May 1983.

STAFF WRITER, "Rapid Prototyping," *System Development*, May 1983.

STAFF WRITER, "Selling Quality," *System Development*, May 1983.

SULLIVAN, LOUIS H., "The Tall Office Building Artistically Considered," *Lippincott's Magazine*, March 1896.

TAKEUCHI, HIROTAKA, and JOHN A. QUELCH, "Quality is More Than Making a Good Product," *Harvard Business Review*, July–August 1983, 139–145. An article that relates surprisingly well to the principles of the structured techniques, but in a nonsystems environment.

TELLEFSEN, GERALD, "Information Technology Management," *Business Week*, March 21, 1983, 61–98.

THADHANI, A. J., "Interactive User Productivity," *IBM Systems Journal*, 20, no. 4 (1981).

THOMSETT, ROB, *People and Project Management* (New York: Yourdon Press, 1980).

TRASS, COLIN G., and LOWELL S. BENSKY, "Systems with Staying Power: A Design Approach to Fault Tolerance," *Computerworld*, February 14, 1983, 11–20.

TUCKER, JONATHAN B., "Computers and the Brain: An A.I. Perspective," *High Technology*, August 1984, 39.

VANDERBOOR, ROMMIE, *An SDM for Maintenance* (MaxiMoney, Inc., Toronto, Canada, January 1985).

VON OECH, ROGER, *A Whack on the Side of the Head: How to Unlock Your Mind for Innovation* (New York: Warner, 1983).

WALTZ, DAVID L., "Artificial Intelligence," *Scientific American*, October 1982, 118–133.

WARD, PAUL T., *Systems Development Without Pain* (New York: Yourdon Press, 1984).

WARD, PAUL T. and STEPHEN J. MELLOR. *Structured Development for Real-Time Systems* (New York: Yourdon Press, 1985).

WARNIER, JEAN-DOMINIQUE, *Logical Construction of Programs*, 3rd ed. (New York: Van Nostrand Reinhold, 1974).

WHITESIDE, DAVID, "Artificial Intelligence: The Race to Make It Work for Managers," *International Management*, September 1984, 20–21.

YOURDON, EDWARD, *Structured Walkthroughs*, 3rd ed. (New York: Yourdon Press, 1985).

———, *The Second Structured Revolution* (New York: Yourdon, Inc., March 1979).

———, *Managing the System Life Cycle* (New York: Yourdon Press, 1982).

YOURDON, EDWARD (ed). *Classics in Software Engineering* (New York: Yourdon Press, 1979).

———, and LARRY L. CONSTANTINE, *Structured Design* (New York: Yourdon Press, 1978).

YUKIO, MIZUNO, "Software Quality Improvement," *Computer*, March 1983, 66–72.

ZELKOWITZ, M. V., "Perspectives on Software Engineering," *ACM Computing Surveys*, 10, no. 2, June 1978.

ZELLS, LOIS, "A Practical Approach to a Project Expectations Document," *Computerworld*, August 29, 1983.

INDEX

A

Acceptance Criteria, 46, 48, 130–32, 146, 155, 157, **211–12**
Adapt, don't adopt, 210
Analysis and design deliverables, 193
Analyst, characteristics of, 6
Approval Group List, 61–62, **212–13**
Asynchronous network model, 29
Audit and security, 120–23
Audit & Security Guidelines, 120, 122, 142, **213–14**
Automation boundary, 45, 64, 102–5, 108, 124, 128, 139

B

Barnes, Djuna, 10
Beer, Stafford, 210
Blitz, 108
Block, Robert, 3, 159
Boar, Bernard H., 184
Boehm, Barry W., 194, 203
Bridge between analysis and design, 136
Brill, Allan, 119, 214
BSP (business system plan), 63
Build and test, 48–50
Business activity patterns, 4
    shown in supermodels, 111

C

Changes to system, 55
Client (The), 17, 23, 32, 36–37, 39, 43, 45–46, 109, 142, 146, 157, **215**
    and the Development Team, 113
    role of, 20, 62, 128
Client List, 61–62, **215**
Client Training Package, 152, 155, **216–17**
Coded Programs, 155, 162, **217**
Constantine, Larry L., 145, 222
Context diagram, 17, 30–34

Controllers:
  role of, 121
Controls (*see also* Audit & Security Guidelines)
  in Process Model, 119
Conversion, 50–52, 152
  costs, 105
  miniature SSDM, 152
  scope, 152, 217–18
Converted Database, 152, **218**
Cost-Benefit Statement, 40, 45, 103–9, 124, 126, 128, **218–20**
  buy a software package option, 107
  develop a new system option, 107, 128, 129
  do nothing option, 106, 127, 129, 207
  example, 132
  immediate opportunities for improvement, 106
  iterations of, 129, 146, 157

D

Data, 32 (*see also* Entity-relationship diagram; Information Model)
  as a corporate resource, 113, 116
Database design, 51, 145, **220**
Data dictionary, 72 (*see also* Chapter 9)
  integration of models, 109
  administrator, 230
Data Dictionary, 147, **221**
Data group, 113
Data model (*see* Information Model)
Data-structured methodologies, 176
DeMarco, Tom, 13, 15, 17, 50, 67, 201, 203, 221, 239, 246, 260
Design, 46–48, 124, 131, 137
Design Blueprint, 48, 56, 82, 147, 162, 196, **222**
Development Team (The), 22–23, 32, 36–37, 39, 40–41, 43, 45–46, 69, 71, 104–5, 109, 120, 128, 145–46, 155, 159, 161, 192–93, 199, **222–23**
  capability influences costs, 194, 209–10
  characteristics of, 66
  composition of, 65, 197
  importance of user participation, 21, 53, 64
  role of, 21
  skills, 60, 157, 186–87, 191
  value of team approach, 194
Dialogue Design, 142, 144, 147, 155, **224**
Dickinson, Brian, 4
Dictionary, 21–22 (*see also* Data Dictionary; Chapter 9)
Dijkstra, E.W., 41, 124
Disaster Recovery Plan, 46, 150, **225**
Documentation, 150–51, 195
  conventional vs effective, 196
  unnecessary, 30, 151, 208
  used in training, 155
Documentation Heaven, 32–33, 67–68, 155, **227–228** (*see also* Stuff Kept and Maintained)
Documentation specialist, 197–98, 217
Document Requirements, **226**
DP security:
  role of, 121, 123

E

EBM (*see* Essential Business Model)
EDP audit (*see* Internal audit)
EIM (*see* Essential Information Model)
Entity-relationship diagram, 75, 90, 91, 113
Environmental Acceptance Criteria, 146, **229–30**
Environment Testing Package, **228–29**
EPM (*see* Essential Process Model)
ERD (*see* Entity-relationship diagram)
Essential Business Model, 21, 36–43, 45, 56, 62, 68, 104, 107, 110, 112, 129, 162, 196, **230**
  creating, 69–72
  normalization of, 93
  physical constraint, 39
  premature assumptions of technology, 108
  productivity, 72
  purpose of, 37

Index

relationship between Essential Information Model and Essential Process Model, 89–92
supermodel, 110
Essential Information Model, 37–38, 63, 69–70, 78–79, 89, 92, 109, 111, 120, **230–31**
Essential Process Model, 37–38, 63, 69–71, 80, 84, 89–92, 96, 103–5, 109, 111, 120, **232–33**
Estimating, 195, 203 (*see* also Cost-Benefit Statement)
by Development Team, 199
Estimating database, 157
Event-based analysis, 11, 72, 183, 201
example of, 73
Event list, 71–73, 77–78
Event modeling, 33
Events, 33
external events, 71
temporal events, 71
Expert systems, 25

F

Feigenbaum, Edward A. and Pamela McCorduck, 147
Fermat's Last Theorem, 23
Flavin, Matt, 70, 231, 236
Frantzen, Trond, 13
FSM (*see* Functional System Model)
Functional System Model, 21, 43–46, 51, 56, 82, 107, 113, 120, 123–24, 128–29, 132, 137, 151, 157, 162, 196, **233–34**
Function Testing Package, **234–35**

G

Game Plan for Systems Development, 10–11, 175
Giraudoux, Jean, 9
Glass, Robert, 150

H

Hacker, 122
Hansen, H. Dines, 51
Hansen, Kirk, 12, 145

Hardware Requirements, **235–36**
Help screens, 155
Hiscock, E. F. Lloyd, 59

I

IBM, 63, 78
Implement, 53–54
Information Model, 113–14, 124, 182, **236–37**
example, 116-19
Integration Testing Package, **237–38**
Internal audit:
role of, 122–23
Intersystem Testing Package, **238–39**
IOI (immediate opportunities for improvement) (*see* Cost-Benefit Statement)

J

Jackson, Michael, 12, 145
JAD (joint application design), 78
Jones, T. Capers, 203

K

Kelly-Bootle, Stan, 204
Knuth, Donald E., 41

L

Lerner, Eric J., 201
Logical model, 38, 68, 72
Lump Law, 208

M

Maintenance, 54–57, 155, 160, 164
costs, 158
need for SuperSet System Development Model in, 158, 162
"SDM for maintenance" example, 164–74
Management support, 179–80
Martin, James, 110
Maximoney, Inc.:
implementation of superset approach, 176

Maximoney, Inc. *(cont.)*
  "SDM for maintenance" example, 164–74
McMenamin, Stephen *(see* Palmer, John)
McMenamin, Stephen and John Palmer, 15, 17, 140
Mellor, Stephen J., 251
Mentors, 189, 191–92, 194, 204
Metrics, 200
  measuring quality, 202
Mini-model, 78, 89, 93
Minsky, Marvin, 95
Models of essence, 17–20 *(see also* Essential Business Model)
  using to solve business problem, 17–20
Module Specifications, 144, 147, **239**
Molnar, Louie A., 150, 243
Myers, Glenford, 150

N

Neighborhood diagram, 34
Newton-John, Sir Isaac, 17

O

Object *(see* Data group)
Objectives & Constraints, 36, 61–62, **240**
  refining, 97
Office Equipment Requirements, **241**
Operations, 150, 155, 241
  role of, 32
Operations Flowchart, 48, 150, **242**
Operations Support Procedures, 50, 196, **243**
Operations Training Package, 152, **244**
Orr, Ken, 12, 68, 145

P

Page-Jones, Meilir, 137, 144, 147, 222, 252

Palmer, John, 15, 214, 232 *(see also* McMenamin, Stephen)
Personnel Requirements, **245**
Physical model, 68, 73, 140
Physiological model, 123
Postproject review, 54, 155, 157
  user participation, 195
Preliminary Agreement, 34–36, 61–62, **245–46**
Problems *(see also* System development methodology–failure of)
  inadequate analysis and design, 180
Process Model, 113, 114, 119, 122, 139, 140, **246**
  example, 124
Processor Configuration Model, 39, 45, 48, 104, 105, 109, 124, 139, 140, **248**
Process specification, 72
Producers and directors, 204–5
Productivity, 200, 206
Project architect, 204
Project life cycle *(see* System development methodology)
Project management and control, 198–203
Project manager, 42, 45, 59, 108, 183, 198–99, 204
  role of, 42
Prototyping, 142, 144, 181–84, 201

Q

Quality indicator formula, 202
Quality Management Team, 157, 159–61, **249**

R

Recursion, 276
Request for Service, 33–34, **350**
Ruskin, John, 181

S

Schuldt, Gary, 66
Scouting blitz, 69

Index

SDM (*see* System development methodology)
Security, 155 (*see also* Audit & Security Guidelines)
  threat from employees, 122
Semprevivo, Philip C., 66, 194, 223
Software:
  buying packaged software, 131
  use of system models in buying, 132
Software repair, 63
Software Support Requirements, **250–51**
SSDM (*see* SuperSet System Development Model)
State Transition Diagrams, 147, **251–52**
Strategic planning, 110
Structure Chart, 144–45, 147, **252**
Structured analysis:
  standardized tools, 189, 192
  tool to develop SSDM, 15
Stuff Kept and Maintained, 30, 33, 57, 197, **252–55**
Sullivan, Louis, 10–11
Superset approach, 164, 198
  at MaxiMoney, Inc., 176
  train users in, 175–76
  usefulness in project management, 198–99
SuperSet System, Development Model, 1–2, 11, 27–30, 34, 60, 209, **255–56**
  adapt, don't adopt, 25, 59
  build using modeling tool, 13, 16, 57, 60,
  customize, 136, 145, 209–10
  iterative, 97, 210
  in maintenance, 162
  must be maintained, 180
  paybacks, 181
  proportion of effort in different phases, 147
  reasons for superset approach, 15
  recreate, 4
Syntelligence, 25
System development:
  a business activity, 7, 62
  automated tools for, 205
  proportion of effort in different phases, 200
System development methodology:
  compared to universal vegetable, 9
  failure of, 3, 7–10, 64
  SuperSet, 3
System in Production, 160, **256**
System models:
  essential, 17
  logical, 17
  need for creativity in building 24, 41
  physical, 17
  text components of, 113–14
System release, 50, 155, 256
System development model:
  deliverables of, 8
  need for, 15, 45
  vs System development methodology, 8, 42
System Support Requirements, 48, 145, **256**

T

Teams (*see* Development Team; testing team)
Tested Conversion System, **259–60**
Test Files, **257**
Testing Team, 46, 131, **260–61**
  independence of, 147
Test Packages, **257**
Test Plan, 48, 146, **258–59**
Things going on in the background, 23–25, 32, 40–42, 142, 157
Thomsett, Rob, 66, 223
Training, 152 (*see also* Client Training Package; Operations Training Package)
  analysts and designers, 186–87
  apprenticeship, 186, 191, 194
  management, 188
  objectives of, 185
  timeliness, 189
  to use SuperSet System Development Model at MaxiMoney, Inc., 178
  users, 188

Training materials, 52–53
Tucker, Johnathan B., 201

U

Universal vegetable, 9
User Documentation, 50, 150–51, 196, **261–63**
Users, 17, 23, 45, 142, 147, 155, 193
 role of, 20, 53, 64, 78, 128, 139, 151
 training, 188
 understanding of superset approach, 175
 use of system models, 129

V

Vanderboor, Rommie, 164
Versioning, 39, 50, 253

Version Plan, 39–40, 45, 51, 103, 124, 152, 263
 used for cost benefit, 104–6

W

Walkthroughs, 60, 193, 210
Ward, Paul T., 4, 15, 33, 113, 231–32, 236, 246, 251
Warnier, J.D., 12, 145
Weinberg, Gerald, 208

Y

Yourdon, Edward, 3, 13, 60, 145, 222

Z

Zelkowitz, M.V., 158
Zells, Lois, 240

READER COMMENT FORM

Your opinions count!

If you have comments, criticisms, or suggestions, we're eager to get them. Your opinions today will affect our products tomorrow. If you have questions, you can expect an answer within two weeks from the time we receive them. And if you discover any errors in this book, typographical or otherwise, please point them out so we can make corrections when the book is reprinted.

Please send your comments and questions directly to us at Benetech Systems, Suite 1805, 77 Bloor Street West, Toronto, Canada M5S 1M2.

Thanks for your help.

Trond Frantzen
Ken McEvoy   Toronto, Canada

Book Title: A GAME PLAN FOR SYSTEMS DEVELOPMENT

Dear Trond and Ken:

_____

_____

_____

_____

_____

_____

Name and Title _____

Company (if any) _____

Address _____

City, State/Province _____

Country _____ Zip/Postal _____

TEAR OUT THIS PAGE TO ORDER THESE OTHER HIGH-QUALITY YOURDON PRESS COMPUTING SERIES TITLES

Quantity	Title/Author	ISBN	Price	Total $
	Building Controls Into Structured Systems; Brill	013-086059-X	$32.00	
	C Notes: Guide to C Programming; Zahn	013-109778-4	$16.95	
	Classics in Software Engineering; Yourdon	013-135179-6	$37.33	
	Controlling Software Projects; DeMarco	013-171711-1	$36.33	
	Creating Effective Sofware; King	013-189242-8	$33.00	
	Crunch Mode; Boddie	013-194960-8	$27.00	
	Current Practices in Software Development; King	013-195678-7	$33.33	
	Data Factory; Roeske	013-196759-2	$22.00	
	Developing Structured Systems; Dickinson	013-205147-8	$32.00	
	Design of On-Line Computer Systems; Yourdon	013-201301-0	$47.00	
	Essential Systems Analysis; McMenamin/Palmer	013-287905-0	$32.00	
	Expert System Technology; Keller	013-295577-6	$26.95	
	Concepts of Information Modeling; Flavin	013-335589-6	$26.67	
	Game Plan for System Development; Frantzen/McEvoy	013-346156-4	$30.00	
	Intuition to Implementation; MacDonald	013-502196-0	$22.00	
	Managing Structured Techniques; Yourdon	013-551037-6	$32.00	
	Managing the System Life Cycle 2/e; Yourdon	013-551045-7	$33.00	
	People & Project Management; Thomsett	013-655747-3	$21.33	
	Politics of Projects; Block	013-685553-9	$21.33	
	Practice of Structured Analysis; Keller	013-693987-2	$26.67	
	Program It Right; Benton/Weekes	013-729005-5	$21.33	
	Software Design: Methods & Techniques; Peters	013-821828-5	$32.00	
	Structured Analysis; Weinberg	013-854414-X	$39.33	
	Structured Analysis & System Specifications; DeMarco	013-854380-1	$42.67	
	Structured Approach to Building Programs: BASIC; Wells	013-854076-4	$21.33	
	Structured Approach to Building Programs: COBOL; Wells	013-854084-5	$21.33	
	Structured Approach to Building Programs: Pascal; Wells	013-851536-0	$21.33	
	Structured Design; Yourdon/Constantine	013-854471-9	$48.00	
	Structured Development Real-Time Systems, Combined; Ward/Mellor	013-854654-1	$70.33	
	Structured Development Real-Time Systems, Vol. I; Ward/Mellor	013-854787-4	$32.00	
	Structured Development Real-Time Systems, Vol. II; Ward/Mellor	013-854795-5	$32.00	
	Structured Development Real-Time Systems, Vol. III; Ward/Mellor	013-854803-X	$32.00	
	Structured Systems Development; Orr	013-855149-9	$32.00	
	Structured Walkthroughs 3/e; Yourdon	013-855248-7	$23.33	
	System Development Without Pain; Ward	013-881392-2	$32.00	
	Teams in Information System Development; Semprivivo	013-896721-0	$26.67	
	Techniques of EDP Project Management; Brill	013-900358-4	$32.00	
	Techniques of Program Structure & Design; Yourdon	013-901702-X	$42.67	
	Up and Running; Hanson	013-937558-9	$28.67	
	Using the Structured Techniques; Weaver	013-940263-2	$25.00	
	Writing of the Revolution; Yourdon	013-970708-5	$37.33	
	Practical Guide to Structured Systems 2/e; Page-Jones	013-690769-5	$35.00	

Total $ _____
- discount (if appropriate) _____
New Total $ _____

OVER PLEASE ➡

AND TAKE ADVANTAGE OF THESE SPECIAL OFFERS!

a.) When ordering 3 or 4 copies (of the same or different titles), take 10% off the total list price (excluding sales tax, where applicable).

b.) When ordering 5 to 20 copies (of the same or different titles), take 15% off the total list price (excluding sales tax, where applicable).

c.) To receive a greater discount when ordering 20 or more copies, call or write:

Special Sales Department
College Marketing
Prentice Hall
Englewood Cliffs, NJ 07632
201-592-2498

SAVE!
If payment accompanies order, plus your state's sales tax where applicable, Prentice Hall pays postage and handling charges. Same return privilege refund guaranteed. Please do not mail in cash.

☐ **PAYMENT ENCLOSED**—shipping and handling to be paid by publisher (please include your state's tax where applicable).

☐ **SEND BOOKS ON 15-DAY TRIAL BASIS** & bill me (with small charge for shipping and handling).

Name _____

Address _____

City _____ State _____ Zip _____

I prefer to charge my   ☐ Visa   ☐ MasterCard
Card Number _____ Expiration Date _____

Signature _____

All prices listed are subject to change without notice.

Mail your order to: Prentice Hall, Book Distribution Center, Route 59 at Brook Hill Drive, West Nyack, NY 10995

Dept. 1   D-OFYP-FW(1)